Contemporary Clinical Neuroscience

Series Editor

Mario Manto, Division of Neurosciences, Department of Neurology, CHU-Charleroi, Charleroi, Belgium, University of Mons, Mons, Belgium, Charleroi, Belgium

Contemporary Clinical Neurosciences bridges the gap between bench research in the neurosciences and clinical neurology work by offering translational research on all aspects of the human brain and behavior with a special emphasis on the understanding, treatment, and eradication of diseases of the human nervous system. These novel, state-of-the-art research volumes present a wide array of preclinical and clinical research programs to a wide spectrum of readers representing the diversity of neuroscience as a discipline. The book series considers proposals from leading scientists and clinicians. The main audiences are basic neuroscientists (neurobiologists, neurochemists, geneticians, experts in behavioral studies, neurophysiologists, neuroanatomists), clinicians (including neurologists, psychiatrists and specialists in neuroimaging) and trainees, graduate students, and PhD students.

Volumes in the series provide in-depth books that focus on neuroimaging, ADHD (attention deficit hyperactivity disorder and other neuropsychiatric disorders, neurodegenerative diseases, G protein receptors, sleep disorders, addiction issues, cerebellar disorders, and neuroimmune diseases. The series aims to expand the topics at the frontiers between basic research and clinical applications. Each volume is available in both print and electronic form.

More information about this series at http://www.springer.com/series/7678

Kenji Hashimoto • Mario Manto
Editors

New Rapid-acting Antidepressants

Springer

Editors
Kenji Hashimoto
Center for Forensic Mental Health
Chiba University
Chiba, Japan

Mario Manto
Service de Neurologie
CHU-Charleroi
Bruxelles, Belgium

ISSN 2627-535X ISSN 2627-5341 (electronic)
Contemporary Clinical Neuroscience
ISBN 978-3-030-79789-8 ISBN 978-3-030-79790-4 (eBook)
https://doi.org/10.1007/978-3-030-79790-4

This Springer imprint is published by the registered company Springer Nature Switzerland AG
The registered company address is: Gewerbestrasse 11, 6330 Cham, Switzerland

Preface

Depression is a major mental health condition globally, with a high rate of relapse. Depression impacts strongly the daily lives of patients. The social and economic burden is considerable. There are important limitations to currently available antidepressants, such as delayed onset from weeks to months and low rates of efficacy. The major discovery that a single dose of ketamine, an NMDAR antagonist, can produce rapid and sustained antidepressant effects in major depressive disorders has opened new research avenues in this area. This book presents the latest advances in pharmaceutical treatments based on rapid antidepressants molecules. Our knowledge of these novel medications is increasing tremendously.

This book aims to provide an updated information on rapid antidepressant molecules. We have gathered specialists in the field with an international top reputation. We are indebted to the contributors. We are also very grateful to the editorial team at Springer for their continuous support and professionalism.

This concise book will be an essential reading for psychiatrists, neuropsychologists, neuroscientists, trainees, and students.

Chiba, Japan — Kenji Hashimoto
Bruxelles, Belgium — Mario Manto

Contents

(*R*)-Ketamine: A New Rapid-Acting Antidepressant

Kai Zhang and Kenji Hashimoto

Abstract (*R*)-ketamine (or arketamine) is the (*R*)-enantiomer of ketamine. (*R*)-ketamine, (*R,S*)-ketamine, and (*S*)-ketamine show different affinities for glutamate, opioid, dopamine, and acetylcholine receptors in the brain. In prior research, the intraperitoneal, intranasal, and intracerebroventricular administration of (*R*)-ketamine produced rapid-acting and long-lasting antidepressant effects in rodent models of depression. In 2016, Zanos et al. demonstrated that the metabolism of (*R,S*)-ketamine to (2*R*,6*R*)-hydroxynorketamine (HNK) is essential for its antidepressant effects. However, our data showed that the metabolism of (*R*)-ketamine to (2*R*,6*R*)-HNK is not essential for the antidepressant effects of (*R*)-ketamine in rodents. Interestingly, (*R*)-ketamine might be free of psychotomimetic side effects and abuse potential in humans and rodents when compared with (*R,S*)-ketamine and (*S*)-ketamine. Although the precise molecular and cellular mechanisms underlying the antidepressant effects of (*R*)-ketamine are unclear, α-amino-3-hydroxy-5-methyl-4-isoxazolepropionic acid receptor activation may have a role in its antidepressant actions. In addition, inflammatory bone markers, extracellular-signal-regulated kinase signaling, and gut microbiota may play a part in the antidepressant effects of (*R*)-ketamine. In this chapter, we discuss the recent findings of the rapid-acting antidepressant candidate (*R*)-ketamine.

Keywords (*R*)-ketamine · (*S*)-ketamine · Opioid · Dopamine · Acetylcholine · (2*R*,6*R*)-HNK · RANKL · ERK · Gut bacteria

K. Zhang
Division of Clinical Neuroscience, Chiba University Center for Forensic Mental Health, Chiba, Japan

Department of Psychiatry, Chaohu Hospital of Anhui Medical University, Hefei, China

K. Hashimoto (✉)
Division of Clinical Neuroscience, Chiba University Center for Forensic Mental Health, Chiba, Japan
e-mail: hashimoto@faculty.chiba-u.jp

K. Hashimoto, M. Manto (eds.), *New Rapid-acting Antidepressants*, Contemporary Clinical Neuroscience,
https://doi.org/10.1007/978-3-030-79790-4_1

Rapid-Acting Antidepressant Ketamine

The *N*-methyl-D-aspartate receptor (NMDAR) antagonist ketamine is the first exemplar of a rapid-acting antidepressant with efficacy for treatment-resistant depression for major depressive disorder (MDD) and bipolar disorder (BD) (Krystal et al. 2019). Further, dissociative symptoms of ketamine may not mediate clinical benefits but may signal adequate target engagement by ketamine. Although we know that ketamine elicits rapid-acting and sustained antidepressant actions in treatment-resistant patients with MDD or BD, the precise molecular and cellular mechanisms underlying these processes remain to be elucidated.

What Is (*R*)-Ketamine?

Ketamine is a racemic mixture of equal amounts of the two enantiomers (*R*)-ketamine (or arketamine) and (*S*)-ketamine (or esketamine) (Hashimoto 2019) (Fig. 1). Similar to (*R*,*S*)-ketamine and (*S*)-ketamine, (*R*)-ketamine is also biologically active. (*R*)-ketamine, (*R*,*S*)-ketamine, and (*S*)-ketamine show different affinities for NMDAR, opioid, dopamine, and acetylcholine receptors (Domino 2010). Relative to (*S*)-ketamine, (*R*)-ketamine possesses a four- to fivefold lower affinity for the phencyclidine site of the NMDAR. In accordance, (*R*)-ketamine is significantly less potent than (*R*,*S*)-ketamine and especially (*S*)-ketamine in terms of anesthetic, analgesic, and sedative–hypnotic effects. (*R*,*S*)-ketamine has weak affinity

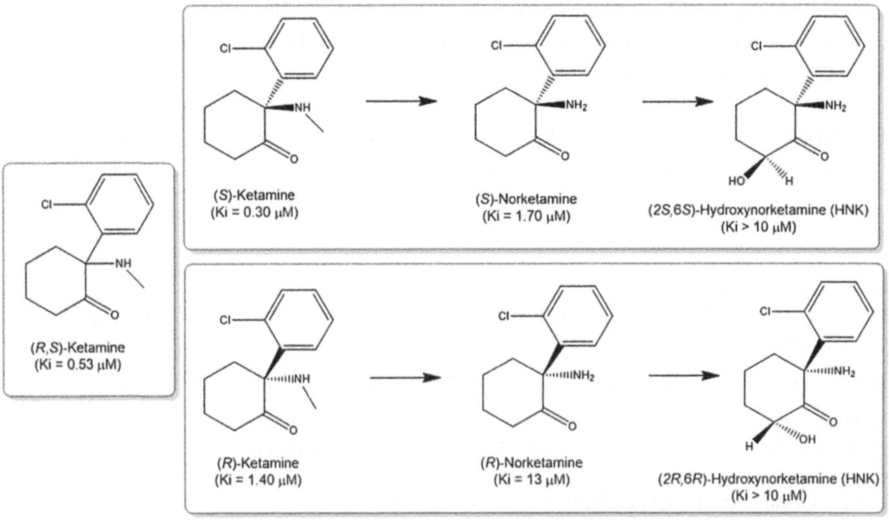

Fig. 1 Chemical structure of enantiomers of ketamine and its metabolites. The values in the parenthesis are the Ki value for the NMDAR (Hashimoto 2019)

for the sigma receptor, where it acts as an agonist, whereas (*S*)-ketamine binds negligibly to this receptor; as such, the sigma receptor activity of (*R,S*)-ketamine lies in (*R*)-ketamine. (*S*)-ketamine inhibits the dopamine transporter about eightfold more potently than does (*R*)-ketamine, and so is about eight times more potent as a dopamine reuptake inhibitor. Both (*R*)-ketamine and (*S*)-ketamine possess similar degrees of potency for interaction with the muscarinic acetylcholine receptors (Domino 2010).

(*R*)-Ketamine's Antidepressant Effects

Zhang et al. (2014) first compared the antidepressant effects of (*R*)-ketamine and (*S*)-ketamine in rodents. A single dose of 10 mg/kg of (*R*)-ketamine or (*S*)-ketamine produced rapid-acting and long-lasting antidepressant effects in juvenile mice exposed to neonatal dexamethasone. Neonatal dexamethasone exposure caused depression-like behaviors in juvenile mice, suggesting that this paradigm may represent a new animal model of pediatric depression (Li et al. 2014). During both a forced swimming test (FST) and a tail suspension test (TST), both enantiomers significantly attenuated the increased immobility time (Zhang et al. 2014). Interestingly, the antidepressant effects of (*R*)-ketamine but not (S)-ketamine could still be detected at 7 days after a single treatment, indicating that (*R*)-ketamine possessed a long-lasting antidepressant effect. This is the first study suggesting an antidepressant effect of (*R*)-ketamine in an animal model of depression. Subsequently, Yang et al. (2015) demonstrated that (*R*)-ketamine has a greater potency and more long-lasting antidepressant effects in comparison with (*S*)-ketamine in a chronic social defeat stress (CSDS) model and learned helplessness (LH) model.

Besides via intravenous infusion, ketamine also can be administered intramuscularly or through other, more convenient means such as orally or intranasally (Zhang and Hashimoto 2019a). Among all the established ketamine administration routes, intramuscular administration has a high bioavailability (93%), with a plasma peak obtained in 5 minutes. The bioavailability rates of intrarectal and intranasal injections are approximately 25% and 50%, respectively (Kronenberg 2002; Hashimoto 2019; Zhang and Hashimoto 2019a).

Intranasal administration is a route of administration in which drugs are insufflated through the nose. It can be a form of either topical administration or systemic administration, as the drugs delivered in this manner can go on to have either purely local or systemic effects. The nasal cavity is covered by a thin mucosa that is well-vascularized. Therefore, a drug molecule can be transferred quickly across the single epithelial cell layer directly to the systemic blood circulation without first-pass hepatic and intestinal metabolism. Chang et al. (2019) compared the antidepressant and side effects of the intranasal administration of (*R*)-ketamine, (*R,S*)-ketamine, and (*S*)-ketamine in mice. The order of antidepressant effects of (*R*)-ketamine, (*R,S*)-ketamine, and (S)-ketamine after intranasal administration in a CSDS model was (*R*)-ketamine > (*R,S*)-ketamine > (*S*)-ketamine. In contrast, the order of side effects [i.e., hyperlocomotion, prepulse inhibition (PPI) deficits, condition place preference (CPP)] in mice was (*S*)-ketamine > (*R,S*)-ketamine > (*R*)-ketamine. These results show that intranasal (*R*)-ketamine is better than (*R,S*)-ketamine and (*S*)-ketamine (Chang et al. 2019).

Brain regions such as the medial prefrontal cortex (mPFC) and hippocampus are known to play a role in the antidepressant effects of ketamine in rodents. Shirayama and Hashimoto (2017a) investigated the effects of a single injection of (*R*)-ketamine into rat brain regions of an LH model of depression. Here, a single bilateral injection of (*R*)-ketamine into the infralimbic portion of the mPFC, CA3, and dentate gyrus (DG) of the hippocampus achieved antidepressant effects in the rat LH model. In contrast, a single bilateral injection of (*R*)-ketamine into the prelimbic portion of the mPFC, shell and core of the nucleus accumbens, basolateral amygdala, and central nucleus of the amygdala did not show antidepressant effects in the rat LH model. Collectively, it is likely that the infralimbic portion of the mPFC, CA3, and DG of the hippocampus might be involved in the antidepressant actions of (*R*)-ketamine (Shirayama and Hashimoto 2017a).

Gender Differences

Some preliminary data have shown that female rats are more sensitive to ketamine's antidepressant effects when compared with male rats (Carrier and Kabbaj 2013; Sarkar and Kabbaj 2016). Zanos et al. (2016) found superior antidepressant-like actions of ketamine in female mice when compared with those in male mice during a FST. However, we did not find the sex-specific differences in the antidepressant actions and pharmacokinetic profiles of (*R*)-ketamine in rodents with a depression-like phenotype (Chang et al. 2018). (*R*)-ketamine significantly attenuated the increased immobility time of FST in lipopolysaccharide (LPS)-treated female and male mice. In addition, there were no sex-specific differences in the concentrations of (*R*)-ketamine in the plasma and brain. These findings indicate there are no sex-specific differences in terms of the acute antidepressant effects and pharmacokinetic profile of (*R*)-ketamine (Chang et al. 2018).

(*R*)-Ketamine and its Final Metabolite, (2*R*,6*R*)-Hydroxynorketamine

Hydroxynorketamine (HNK), or 6-hydroxynorketamine, is a minor metabolite of ketamine. It is formed via hydroxylation of the intermediate, norketamine, which is another major metabolite of ketamine. The major metabolite of ketamine is norketamine (80%). Norketamine is secondarily converted into 4-, 5-, or 6-HNK (15%)—specifically mainly the latter (Singh et al. 2014). Ketamine is also transformed into hydroxyketamine (5%). As such, bioactivated HNK composes less than 15% of a dose of ketamine. In 2016, Zanos et al. (2016) reported that the metabolism of ketamine to (2*R*,6*R*)-HNK is essential for the antidepressant effects of ketamine in rodents. However, our group did not replicate the antidepressant effects of (2*R*,6*R*)-HNK in a CSDS model, an LPS model, or an LH model (Yang et al. 2017b; Shirayama and Hashimoto 2017b). There is now increasing debate ongoing about the antidepressant actions of (2*R*,6*R*)-HNK in rodents (Abdallah 2017; Collingridge et al. 2017; Zanos et al. 2017a; b; Hashimoto and Shirayama 2018; Chaki 2017).

It is well-known that ketamine is rapidly metabolized in the liver by microsomal cytochrome P450 enzymes into norketamine (through *N*-demethylation) and finally into HNK (Ho et al. 2018). To exclude the metabolism of ketamine to HNK in the liver, we examined whether an intracerebroventricular (i.c.v.) infusion of (*R*)-ketamine or its metabolite, (2*R*,6*R*)-HNK, would show antidepressant effects in a CSDS model (Zhang et al. 2018a). We found evidence of (2*R*,6*R*)-HNK in the brain after the i.c.v. infusion of (*R*)-ketamine, although the brain concentration of (2*R*,6*R*)-HNK after the i.c.v. infusion of (*R*)-ketamine was lower than that seen after the i.c.v. infusion of (2*R*,6*R*)-HNK. Furthermore, high concentrations of (2*R*,6*R*)-HNK were detected in the blood and liver after the i.c.v. infusion of (*R*)-ketamine. Nonetheless, the concentration of (2*R*,6*R*)-HNK in the brain after the i.c.v. infusion of (*R*)-ketamine was lower than that after i.c.v. infusion of (2*R*,6*R*)-HNK. In addition, high concentrations of (2*R*,6*R*)-HNK were also found in the blood and liver after the i.c.v. infusion of (2*R*,6*R*)-HNK, indicating the rapid washout into the periphery from the brain. Importantly, an i.c.v. infusion of (*R*)-ketamine but not of (2*R*,6*R*)-HNK produced antidepressant effects in a CSDS model (Zhang et al. 2018a). These data suggest that (*R*)-ketamine in the periphery after washout from the brain is metabolized to (2*R*,6*R*)-HNK in the liver and, subsequently, (2*R*,6*R*)-HNK enters into brain tissues.

Deuterium exchange is a chemical reaction in which a covalently bonded hydrogen atom is replaced by a deuterium atom or vice versa. Deuterium substitution is considered as a potential means of slowing drug metabolism or redirecting sites of metabolism in some cases (Pohl and Gillette 1984). Zanos's group (2016) prepared (*R*,*S*)-6,6-dideutero-ketamine [or (*R*,*S*)-d_2-ketamine] by way of a deuterium substitution at the C6 position. (*R*,*S*)-d_2-ketamine (Ki = 0.833μM for NMDAR) did not change NMDAR's binding affinity or NMDAR-mediated hyperlocomotion. (*R*,*S*)-d_2-ketamine robustly hindered its metabolization to (2*S*,6*S*;2*R*,6*R*)-HNK without changing the (*R*,*S*)-ketamine levels in the brain (Zanos et al. 2016). We examined whether deuterium substitution at the C6 position of (*R*)-ketamine could affect the antidepressant effects of (*R*)-ketamine in a CSDS model (Zhang et al. 2018c). Pharmacokinetic studies showed that levels of (2*R*,6*R*)-d_1-hydroxynorketamine [(2*R*,6*R*)-d_1-HNK], a final metabolite of (*R*)-d_2-ketamine, in the plasma and brain following the administration of (*R*)-d_2-ketamine (10 mg/kg) were lower than those of (2*R*,6*R*)-HNK from (*R*)-ketamine (10 mg/kg), indicating deuterium isotope effects in the production of (2*R*,6*R*)-HNK. In contrast, levels of (*R*)-ketamine and its metabolite (*R*)-norketamine in the plasma and brain were the same for both compounds. In a CSDS model, both (*R*)-ketamine (10 mg/kg) and (*R*)-d_2-ketamine (10 mg/kg) showed rapid and long-lasting (7 days) antidepressant effects, indicating that no deuterium isotope effect existed among the antidepressant effects of (*R*)-ketamine. This study suggests that the deuterium substitution of hydrogen at the C6 position slows the metabolism from (*R*)-ketamine to (2*R*,6*R*)-HNK in mice. In contrast, we did not observe any deuterium isotope effects in terms of the rapid and long-lasting antidepressant effects of (*R*)-ketamine in a CSDS model. Therefore, it is unlikely that (2*R*,6*R*)-HNK is essential for the antidepressant effects of (*R*)-ketamine.

Side Effects

Ketamine has notable side effects, such as psychotomimetic symptoms, abuse potential, and neurotoxicity (Luckenbaugh et al. 2014; Li et al. 2011). The clinical use of ketamine is limited due to its side effects.

A randomized study involving healthy male volunteers ($n = 10$) showed that subjective side effects were more pronounced for (*S*)-ketamine than (*R*)-ketamine (Vollenweider et al. 1997). Elsewhere, a positron-emission tomography study among healthy volunteers demonstrated that psychotomimetic doses of (*S*)-ketamine markedly increased cerebral metabolic rates of glucose in the frontal cortex and thalamus. In contrast, equimolar doses of (*R*)-ketamine tended to decrease cerebral metabolic rates of glucose across the brain, producing no psychotic symptoms but, instead, a state of relaxation and well-being. Thus, it would appear that the psychotomimetic and hyperfrontal metabolic actions of ketamine are mainly induced by (*S*)-ketamine (Hashimoto 2019; Vollenweider et al. 1997; Zanos et al. 2018).

In preclinical studies, behavioral tests such as locomotion, the PPI test, the CPP test, and the novel object recognition test can be performed to measure the side effects of ketamine. (*R*)-ketamine does not appear to cause psychotomimetic effects, based on the lack of behavioral abnormalities observed in rodents after treatment (Yang et al. 2015). In CPP tests, (*R*)-ketamine did not increase CPP scores in mice, whereas (*R,S*)-ketamine and (*S*)-ketamine did so in a significant manner (Yang et al. 2015; Chang et al. 2019). A recent study reported that in the intravenous self-administration model, rats self-administered (*S*)-ketamine but not (*R*)-ketamine (Bonaventura et al. 2021). Combined data points to (*R*)-ketamine being free of psychotomimetic side effects and abuse potential in humans.

Parvalbumin (PV) is a calcium-binding albumin protein that plays a role in many physiological processes such as cell cycle regulation, second messenger production, muscle contraction, the organization of microtubules, and phototransduction (Sohal et al. 2009). Alterations in the function of PV-expressing neurons have been implicated in various areas of clinical diseases, such as Alzheimer's disease, autism, schizophrenia, and age-related cognitive defects (Nakamura et al. 2015). A previous study reported that the loss of PV-positive cells contributes to the pathogenesis of schizophrenia. Repeated treatment with (*R*)-ketamine did not cause a reduction of PV-positive cells in the brain (Yang et al. 2016). However, (*S*)-ketamine induced the loss of PV-positive cells in the same brain region. It seems like that, unlike in the case of (*S*)-ketamine, (*R*)-ketamine did not induce psychotomimetic side effects in humans.

Heat shock proteins (HSPs) are a family of proteins that are produced by cells in response to exposure to stressful conditions (Wang et al. 2004). Many members of this group perform a chaperone function by stabilizing new proteins to ensure correct folding or by helping to refold proteins that were damaged by cell stress. HSP-70, which is known as a sensitive marker of reversible neuronal damage, was induced in the retrosplenial cortices of rat brains after administration of noncompetitive NMDAR antagonists such as dizocilpine, phencyclidine, ketamine, and memantine (Sharp et al. 1995). Tian et al. demonstrated that, similar to dizocilpine and (*R,S*)-ketamine, (*S*)-ketamine (50 and 75 mg/kg) caused marked expression of HSP-70 in the rat retrosplenial cortex (Tian et al. 2018). Unexpectedly, (*R*)-ketamine did not cause the expression of HSP-70 in the same region. These findings suggest that (*R,S*)-ketamine and (*S*)-ketamine have detrimental side effects as compared with (*R*)-ketamine. Given the role of HSP-70 in neuronal injury, it seems that high doses of (*R,S*)-ketamine and (*S*)-ketamine but not (*R*)-ketamine can induce neuropathological changes (e.g., neuronal vacuolization) in the retrosplenial cortices of rat brains.

Furthermore, we reported a significant reduction in the dopamine $D_{2/3}$ receptor binding potential in the monkey striatum after a single intravenous injection of (*S*)-ketamine (0.5 mg/kg for 40-min), but not (*R*)-ketamine (0.5 mg/kg for 40-min) (Hashimoto et al. 2017). This study suggests that (*S*)-ketamine-induced marked dopamine release from presynaptic terminal may contribute to acute side effects (i.e., psychotomimetic and dissociation) in humans (Hashimoto 2019).

Recently, Masaki et al. (2019) demonstrated that (*R*)-ketamine and (*S*)-ketamine produced differential functional magnetic resonance imaging (fMRI) responses in conscious rats. The pharmacological MRI study revealed a significantly positive response to (*S*)-ketamine specifically in the cortex, nucleus accumbens, and striatum. In contrast, negative fMRI responses were observed in various brain regions after (*R*)-ketamine administration. The response provoked by (*S*)-ketamine is similar to that elicited by (*R,S*)-ketamine and dizocilpine, suggesting that NMDAR inhibition might play a role in the response caused by (*S*)-ketamine but not by (*R*)-ketamine (Masaki et al. 2019).

Taken all together, these preclinical and clinical studies suggest that (*R*)-ketamine could be a safer antidepressant than (*S*)-ketamine and (*R,S*)-ketamine (Hashimoto 2014, 2016a, b, c, 2019; Zanos et al. 2018; Yang et al. 2019b).

Potential Mechanisms of Ketamine's Antidepressant Actions

5-Hydroxytryptamine

Serotonin [or 5-hydroxytryptamine (5-HT)] is a monoamine neurotransmitter that plays a role in the actions of antidepressants. It has a popular image as a contributor to feelings of well-being and happiness, though its actual biological function is complex and multifaceted, modulating cognition, reward, learning, memory, and numerous physiological processes (Mohammad-Zadeh et al. 2008). Selective serotonin reuptake inhibitors (SSRIs) can improve depression by increasing the levels of serotonin in the brain (Vaswani et al. 2003). Specifically, SSRIs block the reabsorption (reuptake) of serotonin in the brain, making more serotonin available. SSRIs are labeled as selective because they seem primarily to affect serotonin while avoiding neurotransmitters.

Previous reports have suggested that 5-HT might play a role in the antidepressant actions of ketamine (Gigliucci et al. 2013). We sought to examine whether 5-HT depletion affects the antidepressant actions of (*R*)-ketamine in a CSDS model (Zhang et al. 2017). Para-chlorophenylalanine (PCPA) methyl ester hydrochloride (300 mg/kg, twice daily for 3 consecutive days) or vehicle was administered to control and CSDS-susceptible mice. Ultimately, PCPA treatment caused marked reductions in 5-HT and 5-HIAA in the brain regions of both control and CSDS-susceptible mice. In the TST, (*R*)-ketamine significantly attenuated the increased immobility time in the CSDS-susceptible mice with or without 5-HT depletion. In the sucrose preference test, (*R*)-ketamine significantly enhanced reduced sucrose consumption in the CSDS-susceptible mice with or without 5-HT depletion (Zhang et al. 2017). These findings indicated that 5-HT depletion did not affect the antidepressant effects of (*R*)-ketamine in a CSDS model. Therefore, it is unlikely that 5-HT plays a major role in the antidepressant actions of (*R*)-ketamine.

α-Amino-3-Hydroxy-5-Methyl-4-Isoxazolepropionic Acid Receptor Activation

The α-amino-3-hydroxy-5-methyl-4-isoxazolepropionic acid receptor (AMPAR) is an ionotropic transmembrane receptor for glutamate that mediates fast synaptic transmissions in the central nervous system (Groc et al. 2008). (*R*)-ketamine but not (S)-ketamine significantly reversed the depressive-like behavior induced by repeated treatments with corticosterone in rats at 24 hours after a single administration (Fukumoto et al. 2017). The antidepressant effects of (*R*)-ketamine were attenuated by an AMPAR antagonist, suggesting the involvement of AMPAR stimulation in the effects. The present results confirmed the previous findings that (R)-ketamine and (S)-ketamine exerted rapid and long-lasting antidepressant effects through AMPAR activation (Yang et al. 2015). In contrast, it is unlikely that AMPAR activation plays a role in the antidepressant effects of (*S*)-norketamine, a major metabolite of (*S*)-ketamine, in a CSDS model (Yang et al. 2018a).

Dopamine

Dopamine functions as both a hormone and a neurotransmitter and plays several important roles in the brain and body (Romeo et al. 2018). In the brain, dopamine functions as a neurotransmitter—a chemical released by neurons (nerve cells) to send signals to other nerve cells. It is reported that dopamine $D_{2/3}$ receptors but not dopamine D_1 receptors play a role in the antidepressant actions of (*R*,*S*)-ketamine (Li et al. 2015). A recent study demonstrated that dopamine D_1 receptors in the mPFC of control naïve mice were found to play a role in the antidepressant actions of (*R*,*S*)-ketamine (Hare et al. 2019). However, its role in the antidepressant actions of (*R*)-ketamine, which is more potent than (S)-ketamine, is unknown. In the TST, FST, and 1% sucrose preference test, pretreatment with the dopamine D_1 receptor antagonist SCH-23390 did not block the antidepressant effects of (*R*)-ketamine in susceptible mice after CSDS (Chang et al. 2020). These findings suggest that dopamine D_1 receptors may not play a major role in the antidepressant actions of (*R*)-ketamine although dopamine D_1 receptors may be involved in other pharmacological effects of ketamine (Chang et al. 2020).

Opioid System

Opioids are substances that act on opioid receptors to produce morphine-like effects. Medically, they are primarily used for pain relief, including anesthesia, although other medical uses include the suppression of diarrhea, replacement therapy for opioid use disorder, reversing opioid overdose, suppressing cough, suppressing opioid-induced constipation, and during executions in the United States. Extremely potent opioids such as carfentanil are only approved for veterinary use. Opioids are also frequently used nonmedically for their euphoric effects or to prevent withdrawal.

Williams et al. (2018) demonstrated the role of the opioid system in the rapid antidepressant effects of ketamine in patients with treatment-resistant MDD. In their small-sample, single-center crossover trial, the authors investigated the effects of pretreatment with the opioid receptor antagonist naltrexone (50 mg, 45 minutes before) on the antidepressant effects of ketamine (0.5 mg/kg, intravenous) in the study participants. In ketamine-responsive patients with treatment-resistant depression, pretreatment with naltrexone profoundly attenuated ketamine's antidepressant effects, with none of the ketamine responders meeting the response criterion at Day 1. Further, there were no differences in ketamine-induced dissociation between the two conditions. The authors concluded that opioid receptor activation is required for ketamine's acute antidepressant effects, although the dissociative effects of ketamine are not mediated by the opioid system. However, their sample size ($n = 7$) of ketamine-responsive patients was too small.

We examined whether naltrexone can block the antidepressant effects of ketamine in CSDS and LPS-treated inflammation models of depression (Zhang and Hashimoto 2019b). In the CSDS model, ketamine significantly attenuated the increased immobility time in TST and FST in the CSDS-susceptible mice. Separately, ketamine significantly attenuated the decreased sucrose preference in the 1% sucrose preference test in the CSDS-susceptible mice. However, naltrexone did not block the antidepressant effects of ketamine in either the CSDS model or the LPS-treated model. Naltrexone alone also did not show antidepressant activity in this model. Collectively, these results suggest that opioid receptors do not play a role in the antidepressant effects of ketamine in CSDS and LPS-treated models (Zhang and Hashimoto 2019b).

γ-Aminobutyric Acid

The $GABA_A$ (γ-aminobutyric acid, type A) receptors play a role in a number of psychiatric disorders including depression since the regulation of $GABA_A$ receptors is known to influence glutamate neurotransmission (Kalueff and Nutt 2007; Luscher et al. 2011). Recent studies showed that two negative allosteric modulators (NAMs: L-655,708 and MRK-016) of α5 subunit-containing $GABA_A$ receptors produced rapid antidepressant effects in chronic restraint stress (CRS) and chronic unpredictable stress (CUS) models (Fischell et al. 2015). Unlike (*R,S*)-ketamine, MRK-016 produced no impairment of rota-rod performance, no reduction of prepulse inhibition, no conditioned-place preference, and no change in locomotion (Zanos et al. 2017b). However, there are no reports available at this time showing the comparison of (*R*)-ketamine and these two NAMs in animal models of depression. In addition, there are no reports showing alterations in the expression of α5 $GABA_A$ receptors in susceptible mice or postmortem brain samples from depressed patients. The expression of α5 $GABA_A$ receptors in the prefrontal cortices and hippocampi of CSDS-susceptible mice was significantly higher than that among control mice. Further, the expression of α5 $GABA_A$ receptors in the parietal cortices of depressed patients was also higher than that among control subjects (Xiong et al. 2019b). In the TST and FST, (*R*)-ketamine and MRK-016 significantly attenuated the increased immobility time in susceptible mice as compared with in the vehicle-treated group. In the sucrose preference test, (*R*)-ketamine and MRK-016 significantly enhanced the reduced preference in CSDS-susceptible mice at 2 days after a single injection. However, unlike (*R*)-ketamine, MRK-016 did not attenuate the reduced sucrose preference in susceptible mice at 7 days after a single injection. In contrast, L-655,708 did not show antidepressant effects in the same model. In conclusion, this study shows that increased levels of α5 $GABA_A$ receptors in the PFC and hippocampus may play a role in depression-like phenotypes after CSDS. It is unlikely that MRK-016 has long-lasting antidepressant effects, although it elicits rapid-acting antidepressant effects (Xiong et al. 2018).

Inflammatory Bone Markers

A recent study demonstrated that inflammatory bone markers play a role in the antidepressant functions of (*R,S*)-ketamine in treatment-resistant patients with depression (Kadriu et al. 2018). We examined the effects of inflammatory bone markers in the antidepressant functions of (*R*)-ketamine and (*S*)-ketamine in a CSDS model (Zhang et al. 2018b). Behavioral tests for antidepressant actions were performed after a single administration of (*R*)-ketamine or (*S*)-ketamine. We measured inflammatory bone marker levels in the plasma, including osteoprotegerin, receptor activator of nuclear factor κB ligand (RANKL), and osteopontin. (*R*)-ketamine but not (*S*)-ketamine significantly attenuated increased plasma levels of RANKL in CSDS-susceptible mice. This study also found a positive correlation between sucrose preference and osteoprotegerin/RANKL ratio that may play a role in the antidepressant effects of (*R*)-ketamine.

Subsequently, Xiong et al. (2019a) compared the effects of (*R*)-ketamine and its final metabolite, (2*R*,6*R*)-HNK, in depression-like phenotypes, inflammatory bone markers, and bone mineral density (BMD) in CSDS-susceptible mice. (*R*)-ketamine but not (2*R*,6*R*)-HNK elicited rapid and sustained antidepressant effects in CSDS-susceptible mice. Furthermore, (*R*)-ketamine but not (2*R*,6*R*)-HNK significantly improved the increased plasma levels of RANKL and decreased the OPG/RANKL ratio in CSDS-susceptible mice. Interestingly, (*R*)-ketamine but not (2*R*,6*R*)-HNK significantly attenuated the decreased BMD in CSDS-susceptible mice. These findings demonstrate that (*R*)-ketamine may have beneficial effects in depression-like phenotypes and abnormalities in bone functions of CSDS-susceptible mice. It is, therefore, likely that (*R*)-ketamine could be a potential therapeutic drug for addressing abnormalities in bone metabolism in depressed patients (Zhang et al. 2018b; Xiong et al. 2019a; Wei et al. 2021; Hashimoto 2019).

ERK

The extracellular-signal-regulated kinase (ERK) pathway is one of the major signaling cassettes of the mitogen-activated protein kinase (MAPK) signaling pathway. The ERK cascade is activated by a variety of extracellular agents including growth factors and hormones and also cellular stresses (De Lamirande and Gagnon 2002). A previous study showed the antidepressant effects of fluoxetine via the ERK signaling pathway in a chronic mild stress rat model (First et al. 2011). Yang et al. (2018b) examined whether ERK signaling plays a role in the mechanisms underlying the antidepressant actions of (*R*)-ketamine in a CSDS model of depression. The

ERK inhibitor SL327 significantly blocked the antidepressant effects of (*R*)-ketamine in the CSDS model. Moreover, (*R*)-ketamine significantly attenuated the decreased p-ERK/ERK and p-MEK/MEK ratios in the PFCs and DGs of susceptible mice after CSDS. These findings suggest that ERK signaling plays a role in the antidepressant effects of (*R*)-ketamine (Yang et al. 2018b; Hashimoto 2019).

Gut Microbiota

Multiple lines of evidence indicate that gut microbiota may be involved in the pathophysiology of depression and in the antidepressant-like effects of certain potential candidates (Zhang et al. 2017; Yang et al. 2017a, c, 2019a, b; Wang et al. 2020; Zhang et al. 2019). Recently, the possible role of gut microbiota in the antidepressant effects of (*R*)-ketamine in CSDS-susceptible mice was reported (Qu et al. 2017; Yang et al. 2017c). Furthermore, the antidepressant effects of ketamine might be related to the regulation of the abnormal composition of gut microbiota, while the phylum *Actinobacteria* and the class *Coriobacteriia* might be potential biomarkers for ketamine's antidepressant efficacy (Huang et al. 2019). Cumulatively, it is likely that the antidepressant effects of (*R*)-ketamine and racemic ketamine might be partly mediated by the restoration of altered compositions of the gut microbiota.

Conclusion

Clinical data of non-ketamine NMDAR antagonists (e.g., memantine, traxoprodil, lanicemine, rapastinel, and AV-101) (Hashimoto 2019) and preclinical data using two ketamine enantiomers suggest that mechanisms other than NMDAR inhibition may be involved in the antidepressant effects of ketamine.

On March 5, 2019, the US Food Drug Administration approved an (*S*)-ketamine nasal spray (Spravato™; Janssen Pharmaceutical) for treatment-resistant depression. A clinical study of (*R*)-ketamine and (2*R*,6*R*)-HNK in humans is currently underway (Hashimoto 2019). In the future, it is of great interest to compare the antidepressant effects of (*R*)-ketamine and (*S*)-ketamine [or (2*R*,6*R*)-HNK] in patients with major depressive disorder (Wei et al. 2021).

Acknowledgments This study was supported by AMED (to K.H., JP19dm0107119).

Conflict of Interest Dr. Hashimoto is the inventor of filed patent applications on "The use of R-ketamine in the treatment of psychiatric diseases" and "(S)-norketamine and salt thereof as pharmaceutical" by the Chiba University. Dr. Zhang declares no conflict of interest.

References

Abdallah CG (2017) What's the buzz about hydroxynorketamine? Is it the history, the story, the debate, or the promise? Biol Psychiatry 81(8):e61–e63

Bonaventura J, Lam S, Carlton M, Boehm MA, Gomez JL, Solís O, Sánchez-Soto M, Morris PJ, Fredriksson I, Thomas CJ, Sibley DR, Shaham Y, Zarate CA Jr, Michaelides M (2021) Pharmacological and behavioral divergence of ketamine enantiomers: implications for abuse liability. Mol Psychiatry 2021 Apr 15. https://doi.org/10.1038/s41380-021-01093-2

Carrier N, Kabbaj M (2013) Sex differences in the antidepressant-like effects of ketamine. Neuropharmacology 70:27–34

Chaki S (2017) Is metabolism of (*R*)-ketamine essential for the antidepressant effects? Int J Neuropsychopharmacol 21(2):154–156

Chang L, Toki H, Qu Y, Fujita Y, Mizuno-Yasuhira A, Yamaguchi JI, Chaki S, Hashimoto K (2018) No sex-specific differences in the acute antidepressant actions of (*R*)-ketamine in an inflammation model. Int J Neuropsychopharmacol 21(10):932–937

Chang L, Zhang K, Pu Y, Qu Y, Wang SM, Xiong Z, Ren Q, Dong C, Fujita Y, Hashimoto K (2019) Comparison of antidepressant and side effects in mice after intranasal administration of (*R,S*)-ketamine, (*R*)-ketamine, and (*S*)-ketamine. Pharmacol Biochem Behav 181:53–59

Chang L, Zhang K, Pu Y, Qu Y, Wang SM, Xiong Z, Shirayama Y, Hashimoto K (2020) Lack of dopamine D_1 receptors in the antidepressant actions of (*R*)-ketamine in a chronic social defeat stress model. Eur Arch Psychiatry Clin Neurosci 270(2):271–275

Collingridge GL, Lee Y, Bortolotto ZA, Kang H, Lodge D (2017) Antidepressant actions of ketamine versus hydroxynorketamine. Biol Psychiatry 81(8):e65–e67

De Lamirande E, Gagnon C (2002) The extracellular signal-regulated kinase (ERK) pathway is involved in human sperm function and modulated by the superoxide anion. Mol Hum Reprod 8(2):124–135

Domino EF (2010) Taming the ketamine tiger. 1995. Anesthesiology 113(3):678–684

First M, Gil-Ad I, Taler M, Tarasenko I, Novak N, Weizman A (2011) The effects of fluoxetine treatment in a chronic mild stress rat model on depression-related behavior, brain neurotrophins and ERK expression. J Mol Neurosci 45(2):246

Fischell J, Van Dyke AM, Kvarta MD, LeGates TA, Thompson SM (2015) Rapid antidepressant action and restoration of excitatory synaptic strength after chronic stress by negative modulators of alpha5-containing $GABA_A$ receptors. Neuropsychopharmacology 40(11):2499

Fukumoto K, Toki H, Iijima M, Hashihayata T, Yamaguchi JI, Hashimoto K, Chaki S (2017) Antidepressant potential of (*R*)-ketamine in rodent models: comparison with (S)-ketamine. J Pharmacolog Exper Therapy 361(1):9–16

Gigliucci V, O'Dowd G, Casey S, Egan D, Gibney S, Harkin A (2013) Ketamine elicits sustained antidepressant-like activity via a serotonin-dependent mechanism. Psychopharmacology 228(1):157–166

Groc L, Choquet D, Chaouloff F (2008) The stress hormone corticosterone conditions AMPAR surface trafficking and synaptic potentiation. Nat Neurosci 11(8):868

Hare BD, Shinohara R, Liu RJ, Pothula S, DiLeone RJ, Duman RS (2019) Optogenetic stimulation of medial prefrontal cortex Drd1 neurons produces rapid and long-lasting antidepressant effects. Nat Commun 10(1):223

Hashimoto K (2014) The *R*-stereoisomer of ketamine as an alternative for ketamine for treatment-resistant major depression. Clin Psychopharmacol Neurosci 12:72–73

Hashimoto K (2016a) *R*-ketamine: a rapid-onset and sustained antidepressant without risk of brain toxicity. Psychol Med 46:2449–2451

Hashimoto K (2016b) Ketamine's antidepressant action: beyond NMDA receptor inhibition. Expert Opin Ther Targets 20:1389–1392

Hashimoto K (2016c) Detrimental side effects of repeated ketamine infusions in the brain. Am J Psychiatr 173:1044–1045

Hashimoto K (2019) Rapid-acting antidepressant ketamine, its metabolites and other candidates: A historical overview and future perspective. Psychiat Clin Neurosci 73(10):613–627

Hashimoto K, Kakiuchi T, Ohba H, Nishiyama S, Tsukada H (2017) Reduction of dopamine $D_{2/3}$ receptor binding in the striatum after a single administration of esketamine, but not *R*-ketamine: a PET study in conscious monkeys. Eur Arch Psychiatry Clin Neurosci 267(2):173–176

Hashimoto K, Shirayama Y (2018) What are the causes for discrepancies of antidepressant actions of (2*R*,6*R*)-hydroxynorketamine? Biol Psychiatry 84(1):e7–e8

Ho MF, Correia C, Ingle JN, Kaddurah-Daouk R, Wang L, Kaufmann SH, Weinshilboum RM (2018) Ketamine and ketamine metabolites as novel estrogen receptor ligands: induction of cytochrome P450 and AMPA glutamate receptor gene expression. Biochem Pharmacol 152:279–292

Huang N, Hua D, Zhan G, Li S, Zhu B, Jiang R, Yang L, Bi J, Xu H, Hashimoto K, Luo A, Yang C (2019) Role of Actinobacteria and Coriobacteriia in the antidepressant effects of ketamine in an inflammation model of depression. Pharmacol Biochem Behav 176:93–100

Kadriu B, Gold PW, Luckenbaugh DA, Lener MS, Ballard ED, Niciu MJ, Henter ID, Park LT, De Sousa RT, Yuan P, Machado-Vieira R, Zarate CA (2018) Acute ketamine administration corrects abnormal inflammatory bone markers in major depressive disorder. Mol Psychiatry 23(7):1626–1631

Kalueff AV, Nutt DJ (2007) Role of GABA in anxiety and depression. Depress Anxiety 24(7):495–517

Kronenberg RH (2002) Ketamine as an analgesic: parenteral, oral, rectal, subcutaneous, transdermal and intranasal administration. J Pain Palliat Care Pharmacother 16(3):27–35

Krystal JH, Abdallah CG, Sanacora G, Charney DS, Duman RS (2019) Ketamine: a paradigm shift for depression research and treatment. Neuron 101(5):774–778

Li JH, Vicknasingam B, Cheung YW, Zhou W, Nurhidayat AW, Des Jarlais DC, Schottenfeld R (2011) To use or not to use: an update on licit and illicit ketamine use. Subst Abus Rehabil 2:11

Li SX, Fujita Y, Zhang JC, Ren Q, Ishima T, Wu J, Hashimoto K (2014) Role of the NMDA receptor in cognitive deficits, anxiety and depressive-like behavior in juvenile and adult mice after neonatal dexamethasone exposure. Neurobiol Dis 62:124–134

Li Y, Zhu ZR, Ou BC, Wang YQ, Tan ZB, Deng CM, Gao YY, Tang M, So JH, Mu YL, Zhang LQ (2015) Dopamine D_2/D_3 but not dopamine D_1 receptors are involved in the rapid antidepressant-like effects of ketamine in the forced swim test. Behav Brain Res 279:100–105

Luckenbaugh DA, Niciu MJ, Ionescu DF, Nolan NM, Richards EM, Brutsche NE, Guevara S, Zarate CA (2014) Do the dissociative side effects of ketamine mediate its antidepressant effects? J Affect Disord 159:56–61

Luscher B, Shen Q, Sahir N (2011) The GABAergic deficit hypothesis of major depressive disorder. Mol Psychiatry 16(4):383–406

Masaki Y, Kashiwagi Y, Watabe H, Abe K (2019) (*R*)- and (*S*)-ketamine induce differential fMRI responses in conscious rats. Synapse 73(12):e22126

Mohammad-Zadeh LF, Moses L, Gwaltney-Brant SM (2008) Serotonin: a review. J Vet Pharmacol Ther 31(3):187–199

Nakamura T, Matsumoto J, Takamura Y, Ishii Y, Sasahara M, Ono T, Nishijo H (2015) Relationships among parvalbumin-immunoreactive neuron density, phase-locked gamma oscillations, and autistic/schizophrenic symptoms in PDGFR-β knock-out and control mice. PLoS One 10(3):e0119258

Pohl LR, Gillette JR (1984) Determination of toxic pathways of metabolism by deuterium substitution. Drug Metab Rev 15(7):1335–1351

Qu Y, Yang C, Ren Q, Ma M, Dong C, Hashimoto K (2017) Comparison of (*R*)-ketamine and lanicemine on depression-like phenotype and abnormal composition of gut microbiota in a social defeat stress model. Sci Rep 7(1):15725

Romeo B, Blecha L, Locatelli K, Benyamina A, Martelli C (2018) Meta-analysis and review of dopamine agonists in acute episodes of mood disorder: efficacy and safety. J Psychopharmacol 32(4):385–396

Sarkar A, Kabbaj M (2016) Sex differences in effects of ketamine on behavior, spine density, and synaptic proteins in socially isolated rats. Biol Psychiatry 80(6):448–456
Sharp JW, Petersen DL, Langford MT (1995) DNQX inhibits phencyclidine (PCP) and ketamine induction of the hsp70 heat shock gene in the rat cingulate and retrosplenial cortex. Brain Res 687(1–2):114–124
Shirayama Y, Hashimoto K (2017a) Effects of a single bilateral infusion of *R*-ketamine in the rat brain regions of a learned helplessness model of depression. Eur Arch Psychiatry Clin Neurosci 267(2):177–182
Shirayama Y, Hashimoto K (2017b) Lack of antidepressant effects of (2*R*,6*R*)-hydroxynorketamine in a rat learned helplessness model: comparison with (*R*)-ketamine. Int J Neuropsychopharmacol 21(1):84–88
Singh NS, Zarate CA Jr, Moaddel R, Bernier M, Wainer IW (2014) What is hydroxynorketamine and what can it bring to neurotherapeutics? Expert Rev Neurother 14(11):1239–1242
Sohal VS, Zhang F, Yizhar O, Deisseroth K (2009) Parvalbumin neurons and gamma rhythms enhance cortical circuit performance. Nature 459(7247):698–702
Tian Z, Dong C, Fujita A, Fujita Y, Hashimoto K (2018) Expression of heat shock protein HSP-70 in the retrosplenial cortex of rat brain after administration of (*R,S*)-ketamine and (*S*)-ketamine, but not (*R*)-ketamine. Pharmacol Biochem Behav 172:17–21
Vaswani M, Linda FK, Ramesh S (2003) Role of selective serotonin reuptake inhibitors in psychiatric disorders: a comprehensive review. Prog Neuro-Psychopharmacol Biol Psychiatry 27(1):85–102
Vollenweider FX, Leenders KL, Øye I, Hell D, Angst J (1997) Differential psychopathology and patterns of cerebral glucose utilisation produced by (*S*)-and (*R*)-ketamine in healthy volunteers using positron emission tomography (PET). Eur Neuropsychopharmacol 7(1):25–38
Wang S, Qu Y, Chang L, Pu Y, Zhang K, Hashimoto K (2020) Antibiotic-induced microbiome depletion is associated with resilience in mice after chronic social defeat stress. J Affect Disord 260:448–457
Wang W, Vinocur B, Shoseyov O, Altman A (2004) Role of plant heat-shock proteins and molecular chaperones in the abiotic stress response. Trends Plant Sci 9(5):244–252
Wei Y, Chang L, Hashimoto K (2021) Molecular mechanisms underlying the antidepressant actions of arketamine: beyond the NMDA receptor. Mol Psychiatry 2021 May 7. https://doi.org/10.1038/s41380-021-01121-1
Williams NR, Heifets BD, Blasey C, Sudheimer K, Pannu J, Pankow H, Hawkins J, Birnbaum J, Lyons DM, Rodriguez CI, Schatzberg AF (2018) Attenuation of antidepressant effects of ketamine by opioid receptor antagonism. Am J Psychiatr 175(12):1205–1215
Xiong Z, Fujita Y, Zhang K, Pu Y, Chang L, Ma M, Chen J, Hashimoto K (2019a) Beneficial effects of (*R*)-ketamine, but not its metabolite (2*R*,6*R*)-hydroxynorketamine, in the depression-like phenotype, inflammatory bone markers, and bone mineral density in a chronic social defeat stress model. Behav Brain Res 368:111904
Xiong Z, Zhang K, Ren Q, Chang L, Chen J, hashimoto K (2019b) Increased expression of inwardly rectifying Kir4.1 channel in the parietal cortex from patients with major depressive disorder. J Affect Disord 245:265–269.
Xiong Z, Zhang K, Ishima T, Ren Q, Chang L, Chen J, Hashimoto K (2018) Comparison of rapid and long-lasting antidepressant effects of negative modulators of α5-containing $GABA_A$ receptors and (*R*)-ketamine in a chronic social defeat stress model. Pharmacol Biochem Behav 175:139–145
Yang C, Shirayama Y, Zhang JC, Ren Q, Yao W, Ma M, Dong C, Hashimoto K (2015) *R*-ketamine: a rapid-onset and sustained antidepressant without psychotomimetic side effects. Transl Psychiatry 5(9):e632
Yang C, Han M, Zhang JC, Ren Q, Hashimoto K (2016) Loss of parvalbumin-immunoreactivity in mouse brain regions after repeated intermittent administration of esketamine, but not *R*-ketamine. Psychiatry Res 239:281–283
Yang C, Fujita Y, Ren Q, Ma M, Dong C, Hashimoto K (2017a) *Bifidobacterium* in the gut microbiota confer resilience to chronic social defeat stress in mice. Sci Rep 7:45942

Yang C, Qu Y, Abe M, Nozawa D, Chaki S, Hashimoto K (2017b) (*R*)-ketamine shows greater potency and longer lasting antidepressant effects than its metabolite (2*R*,6*R*)-hydroxynorketamine. Biol Psychiatry 82(5):e43–e44

Yang C, Qu Y, Fujita Y, Ren Q, Ma M, Dong C, Hashimoto K (2017c) Possible role of the gut microbiota–brain axis in the antidepressant effects of (*R*)-ketamine in a social defeat stress model. Transl Psychiatry 7(12):1294

Yang C, Kobayashi S, Nakao K, Dong C, Han M, Qu Y, Ren Q, Zhang JC, Ma M, Toki H, Yamaguchi JI, Chaki S, Shirayama Y, Nakazawa K, Manabe T, Hashimoto K (2018a) AMPA receptor activation–independent antidepressant actions of ketamine metabolite (*S*)-norketamine. Biol Psychiatry 84(8):591–600

Yang C, Ren Q, Qu Y, Zhang JC, Ma M, Dong C, Hashimoto K (2018b) Mechanistic target of rapamycin–independent antidepressant effects of (*R*)-ketamine in a social defeat stress model. Biol Psychiatry 83(1):18–28

Yang C, Fang X, Zhan G, Huang N, Li S, Bi J, Jiang R, Yang L, Miao L, Zhu B, Luo A, Hashimoto K (2019a) Key role of gut microbiota in anhedonia-like phenotype in rodents with neuropathic pain. Transl Psychiatry 9(1):57

Yang C, Yang J, Luo A, Hashimoto K (2019b) Molecular and cellular mechanisms underlying the antidepressant effects of ketamine enantiomers and its metabolites. Transl Psychiatry 9(1):280

Zanos P, Moaddel R, Morris PJ, Georgiou P, Fischell J, Elmer GI, Alkondon M, Yuan P, Pribut HJ, Singh NS, Dossou KS, Fang Y, Huang XP, Mayo CL, Wainer IW, Albuquerque EX, Thompson SM, Thomas CJ, Zarate CA Jr, Gould TD (2016) NMDAR inhibition-independent antidepressant actions of ketamine metabolites. Nature 533(7604):481–486

Zanos P, Moaddel R, Morris PJ, Wainer IW, Albuquerque EX, Thompson SM, Thomas CJ, Zarate CA, Gould TD (2017a) Reply to: antidepressant actions of ketamine versus hydroxynorketamine. Biol Psychiatry 81(8):e69–e71

Zanos P, Moaddel R, Morris PJ, Riggs LM, Highland JN, Georgiou P, Pereira EFR, Albuquerque EX, Thomas CJ, Zarate CA, Gould TD (2018) Ketamine and ketamine metabolite pharmacology: insights into therapeutic mechanisms. Pharmacol Rev 70(3):621–660

Zanos P, Nelson ME, Highland JN, Krimmel SR, Georgiou P, Gould TD, Thompson SM (2017b) A negative allosteric modulator for α5 subunit-containing GABA receptors exerts a rapid and persistent antidepressant-like action without the side effects of the NMDA receptor antagonist ketamine in mice. Eneuro 4(1):e0285–e0216

Zhang K, Dong C, Fujita Y, Fujita A, Hashimoto K (2017) 5-Hydroxytryptamine-independent antidepressant actions of (*R*)-ketamine in a chronic social defeat stress model. Int J Neuropsychopharmacol 21(2):157–163

Zhang K, Fujita Y, Chang L, Qu Y, Pu Y, Wang S, Shirayama Y, Hashimoto K (2019) Abnormal composition of gut microbiota is associated with resilience versus susceptibility to inescapable electric stress. Transl Psychiatry 9:231

Zhang K, Fujita Y, Hashimoto K (2018a) Lack of metabolism in (*R*)-ketamine's antidepressant actions in a chronic social defeat stress model. Sci Rep 8(1):4007

Zhang K, Ma M, Dong C, Hashimoto K (2018b) Role of inflammatory bone markers in the antidepressant actions of (*R*)-ketamine in a chronic social defeat stress model. Int J Neuropsychopharmacol 21(11):1025–1030

Zhang K, Toki H, Fujita Y, Ma M, Chang L, Qu Y, Harada S, Nemoto T, Mizuno-Yashira A, Yamaguchim JI, Chaki S, Hashimoto K (2018c) Lack of deuterium isotope effects in the antidepressant effects of (*R*)-ketamine in a chronic social defeat stress model. Psychopharmacology 235(11):3177–3185

Zhang K, Hashimoto K (2019a) An update on ketamine and its two enantiomers as rapid-acting antidepressants. Expert Rev Neurother 19(1):83–92

Zhang K, Hashimoto K (2019b) Lack of opioid system in the antidepressant actions of ketamine. Biol Psychiatry 85(6):e25–e27

Zhang JC, Li SX, Hashimoto K (2014) *R*(−)-ketamine shows greater potency and longer lasting antidepressant effects than *S*(+)-ketamine. Pharmacol Biochem Behav 116:137–141

(2R,6R)-Hydroxynorketamine, A Metabolite of Ketamine: The Antidepressant Actions and the Mechanisms

Kenichi Fukumoto and Ronald S. Duman

Abstract Current antidepressant medications acting on monoaminergic systems take long time (weeks to months) to induce beneficial effects at low rates (about 30%). Accumulating evidences demonstrate that ketamine, a noncompetitive N-methyl-D-aspartate (NMDA) receptor antagonist, produces rapid and long-lasting antidepressant effects in patients with major depressive disorder as well as treatment-resistant depression. A metabolite of ketamine, (2R,6R)-hydroxynorketamine [(2R,6R)-HNK], produces rapid and sustained antidepressant-like effects in animal models without ketamine-like side effects, which have recently gained much attention. Notably, (2R,6R)-HNK does not block the NMDA receptor, the primary target of ketamine. However, the molecular signaling mechanisms of antidepressant-like effects of (2R,6R)-HNK remain unknown.

In this chapter, we discuss the rapid and sustained antidepressant-like actions of (2R,6R)-HNK and those roles in ketamine's antidepressant effects. Furthermore, we review the potential mechanisms of antidepressant-like effects of (2R,6R)-HNK.

Keywords Antidepressant · AMPA receptor · BDNF · Depression · Ketamine · mTORC1 · NMDA receptor · (2R,6R)-Hydroxynorketamine

---

K. Fukumoto (✉) ·
Departments of Psychiatry and Neurosciences, Yale University School of Medicine, New Haven, CT, USA

Pharmacology Research Unit, Sumitomo Dainippon Pharma, Kasugade-naka, Konohana-ku, Osaka, Japan
e-mail: kenichi-fukumoto@ds-pharma.co.jp

R. S. Duman
Departments of Psychiatry and Neurosciences, University School of Medicine, New Haven, CT, USA

K. Hashimoto, M. Manto (eds.), *New Rapid-acting Antidepressants*, Contemporary Clinical Neuroscience,
https://doi.org/10.1007/978-3-030-79790-4_2

Introduction

Major depressive disorder (MDD) is a widespread illness with an estimated lifetime prevalence in the United States of ~17% and has high rates of relapse, which is one of major social and economic disabilities worldwide (Kessler et al. 2003; Mueller et al. 1999; Hasin et al. 2018). Although the majority of individuals with depression (about 65%) exhibit some improvement with currently available antidepressant medications which regulate monoamine neurotransmission, these agents can take weeks to months to exert a therapeutic response, and more than 30% of patients fail to respond to these agents and are considered treatment resistant (John Rush et al. 2006). These findings highlight a significant unmet need for greater efficacious and rapid-acting antidepressants with mechanisms distinct from the currently available antidepressants.

Accumulating evidence has demonstrated that a single subanesthetic dose of ketamine, a noncompetitive N-methyl-d-aspartate (NMDA) receptor antagonist, produces efficacious, rapid (within hours), and long-lasting (7 to 10 days) antidepressant effects in MDD patients, including those considered treatment resistant (Aan Het Rot et al. 2012; Berman et al. 2000; Zarate Jr et al. 2006a). Furthermore, ketamine has been reported to show rapid and sustained reductions of suicidal ideation in patients with depression (Ionescu et al. 2016). The discovery of the rapid, sustained, and potent antidepressant effects of ketamine is regarded as the greatest breakthrough in the field of psychiatry in the last two decades. However, ketamine also has undesirable side effects, including cognitive impairment, psychotomimetic and dissociative symptoms, as well as abuse potential (Krystal et al. 2013). The discovery of ketamine triggered research activities to elucidate the mechanisms of antidepressant effects of ketamine and develop other glutamatergic agents exerting the robust antidepressant actions without the side effects of ketamine. Other glutamatergic agents including antagonists of metabotropic glutamate 2/3 receptors as well as NMDA-GluN2B receptor subtype negative allosteric modulators have been particularly focused. On the other hand, other nonselective NMDA receptor antagonists such as lanicemine and memantine have failed to produce the potent antidepressant effects in clinical trials (Sanacora et al. 2014; Zarate Jr et al. 2006b; Hellweg et al. 2012). Although the exact mechanisms of the robust antidepressant effects of ketamine remain unclear, there are subtle differences in the trapping and channel blocking activity for NMDA receptor between these agents and ketamine.

In addition to those glutamatergic agents, a recent study demonstrates that (2R,6R)-hydroxynorketamine [(2R,6R)-HNK], a major metabolite of ketamine, produces rapid and sustained antidepressant-like effects in rodent models without the undesirable ketamine-like side effects including psychotomimetic-like behaviors and abuse potential, which may play a primary role in the therapeutic actions of ketamine (Zanos et al. 2016). Notably, (2R,6R)-HNK does not block the NMDA receptor, the primary target of ketamine (Zanos et al. 2016). This study also suggests that antidepressant-like effects of ketamine are not mediated by blockade of NMDA receptor as previously assumed, which have attracted much attention.

However, the molecular signaling mechanisms of antidepressant-like effects of (2R,6R)-HNK remain unknown. In addition, subsequent other studies and commentaries indicated controversial findings and doubt the hypothesis (Hashimoto 2016; Suzuki et al. 2017; Yang et al. 2017; Collingridge et al. 2017).

In this chapter, we review the rapid and sustained antidepressant-like actions of (2R,6R)-HNK and those potential molecular and cellular mechanisms. Moreover, we discuss the roles of (2R,6R)-HNK in ketamine's antidepressant effects.

Antidepressant-like Actions of (2R,6R)-Hydroxynorketamine

Ketamine, a racemic mixture consisting of equal parts of (R)-ketamine and (S)-ketamine, is extensively and stereoselectively metabolized by cytochrome P450 (CYP) enzymes to norketamine via N-demethylation in the liver initially (Zanos et al. 2018). Subsequently, norketamine is further metabolized to several metabolites, including (2R,6R;2S,6S)-HNK and (2S,6R; 2R,6S)-HNK which were found in human plasma and plasma and brain of mice following ketamine treatment (Zanos et al. 2016, 2018; Zarate Jr et al. 2012).

Zanos et al. (2016) demonstrated that deuterated ketamine, which did not readily metabolize to (2S,6S;2R,6R)-HNK and retained comparable properties of ketamine in the affinity for NMDA receptor and the pharmacokinetics, did not exert ketamine-like sustained antidepressant-like actions in rodents. They also reported that (2R,6R)-HNK exerted ketamine-like antidepressant effects in rodents, and those actions of (2R,6R)-HNK are more potent than (2S,6S)-HNK (Zanos et al. 2016). These findings suggest that (2R,6R)-HNK is responsible for the antidepressant effects of ketamine. (2R,6R)-HNK is reported to induce the antidepressant-like effects in animal models of depression such as chronic social defeat stress (CSDS), learned helplessness (LH), and chronic corticosterone treatment by using forced swimming test, novelty-suppressed feeding test, sucrose preference test, and female urine sniffing test (Zanos et al. 2016), implying that (2R,6R)-HNK have antidepressant effects like ketamine. There are several studies supporting this finding reported by Zanos et al. (2016). (2R,6R)-HNK displayed antidepressant-like effects in naive mice (Pham et al. 2018) and improved inescapable shock-induced depression-like behavior with rapid-acting and long-lasting antidepressant-like effects in modified learned helplessness paradigm in rats (Chou et al. 2018). We also demonstrated that (2R,6R)-HNK induced rapid and long-lasting antidepressant-like actions in behavioral tests in socially isolated mice (Fukumoto et al. 2019). Notably, in contrast to ketamine, (2R,6R)-HNK did not have the ketamine-related side effects such as psychotomimetic actions, abuse potential, as well as sensory-dissociation properties in preclinical tests (Zanos et al. 2016). These findings indicate that (2R,6R)-HNK may be a new promising antidepressant with rapid and long-lasting therapeutic actions and safe profiles.

However, there are some controversial reports about the antidepressant-like effects of (2R,6R)-HNK. Previous studies reported that (2R,6R)-HNK did not exert antidepressant-like effects in CSDS and lipopolysaccharide (LPS)-induced models, although (R)-ketamine, parent compound of (2R,6R)-HNK, elicited robust antidepressant-like actions in those same animal models of depression by using behavioral tests such as tail suspension test, forced swimming test, and sucrose preference test (Yang et al. 2017; Shirayama and Hashimoto 2018; Zhang et al. 2018a, b). Recent study also showed that (2R,6R)-HNK did not exert the antidepressant-like effects in chronic corticosterone-induced model of depression as well as control naïve mice by using forced swimming test and female encounter test (Yokoyama et al. 2020). In addition, microinjection of (R)-ketamine into the brain regions, such as the infralimbic region of the medial prefrontal cortex (mPFC), CA3, and the dentate gyrus of the hippocampus, induced antidepressant-like effects in LH model, suggesting that (R)-ketamine itself exerts the actions (Shirayama and Hashimoto 2017). These findings doubt that (2R,6R)-HNK plays a key role in the antidepressant effects of ketamine.

In respect of pharmacokinetic considerations, there is a notable report to compare the concentrations of (2R,6R)-HNK in the cerebrospinal fluid (CSF), indicating the free fraction of the drug involved in its actions on the brain, after administration of the same dose of (2R,6R)-HNK and (R)-ketamine (Yamaguchi et al. 2018). The levels of (2R,6R)-HNK in the plasma, brain, and CSF at the time of the plasma maximum concentration after (2R,6R)-HNK dosing in rodents are reported to be approximately 20 times higher than those of (2R,6R)-HNK after (R)-ketamine treatment at the same dose (Yamaguchi et al. 2018). Both ketamine and its enantiomer, (R)-ketamine, at same extent dose are reported to produce the antidepressant-like effects in rodents (Fukumoto et al. 2017). In the case that (2R,6R)-HNK is responsible for the antidepressant effects of ketamine or (R)-ketamine, it should need to exert the actions at a dose approximately 20 times lower than that of ketamine or (R)-ketamine. However, there is no report that (2R,6R)-HNK at such a low dose elicits the antidepressant-like effects. Furthermore, although CYP inhibitors prior to (R)-ketamine treatment almost completely blocked the generation of (2R,6R)-HNK, (R)-ketamine produced the antidepressant-like actions at a lower dose in the presence of CYP inhibitors than in their absence, indicating that the antidepressant-like effects of (R)-ketamine depend on exposure levels of (R)-ketamine but not (2R,6R)-HNK (Yamaguchi et al. 2018). Both deuterated (R)-ketamine, which was similar to (R)-ketamine in binding to and functionally inhibiting NMDA receptors but attenuated (R)-ketamine's metabolism to (2R,6R)-HNK, and (R)-ketamine also reported to exert similar rapid and long-lasting antidepressant-like effects in CSDS model in mice (Zhang et al. 2018b). Considering that (2R,6R)-HNK is produced only from (R)-ketamine, these findings suggest that (2R,6R)-HNK is not essential for ketamine as well as (R)-ketamine to exert those antidepressant-like effects. In addition, a recent study reported that deuterated (R)-ketamine also elicited less antidepressant-like actions than (R)-ketamine in rodents, suggesting that (2R,6R)-HNK is partly involved in the antidepressant-like actions of (R)-ketamine (Zanos et al. 2019a).

Collectively, even though (2R, 6R)-HNK has the antidepressant-like potential in rodents, its antidepressant-like profile does not play a main role in the antidepressant actions of ketamine and may be substantially different from those of ketamine.

Brain Regions for the Antidepressant-like Actions of (2R,6R)-Hydroxynorketamine

We reported that microinjection of (2R,6R)-HNK into the mPFC was sufficient to exert antidepressant-like actions similar to systemic treatment of (2R,6R)-HNK (Fukumoto et al. 2019). This finding is consistent with previous reports that the mPFC is sufficient and necessary for the antidepressant-like effects of ketamine as well as the effects of other rapid-acting agents, including rapastinel and scopolamine (Fuchikami et al. 2015; Gerhard et al. 2020; Burgdorf et al. 2013; Navarria et al. 2015). A recent study reported that intra-mPFC infusion as well as systemic administration of (2R,6R)-HNK produced antidepressant-like effects in rodents (Pham et al. 2018). Moreover brain imaging and postmortem studies show the anatomical and functional abnormalities in mPFC in MDD patients and the possibility that mPFC plays important roles in the antidepressant effects of ketamine, implicating the mPFC in both the pathophysiology and treatment of depression. (Holmes and Wellman 2009; Murrough 2012; Price and Drevets 2010). Collectively, these findings provide strong evidence that the mPFC is a key region for the antidepressant-like actions of (2R,6R)-HNK. However, a recent study reported that different with ketamine, (2R,6R)-HNK had no effect on GCaMP6f activity, a genetically encoded calcium indicator representing neuronal activity, in pyramidal cells of ventral mPFC (Hare et al. 2020). This finding suggests that ventral mPFC is not a site of action for (2R,6R)-HNK to elicit the antidepressant-like actions and the initial mechanisms of the actions of (2R,6R)-HNK differ from those of ketamine. Although microinjection of (2R,6R)-HNK into the ventrolateral periaqueductal gray (vlPAG) is reported to exhibit a rapid-acting and long-lasting antidepressant-like effects in LH model, indicating the importance of vlPAG in (2R,6R)-HNK's antidepressant-like actions (Chou et al. 2018), studies of additional areas, including the hippocampus and amygdala, are needed to further characterize the regional effects of (2R,6R)-HNK.

Potential Mechanisms of Antidepressant-like Actions of (2R,6R)-Hydroxynorketamine

Zanos et al. (2016) reported that the rapid and long-lasting antidepressant-like effects of (2R,6R)-HNK were blocked by the AMPA (α-amino-3-hydroxy-5--methyl-4-isoxazole propionic acid) receptor antagonist NBQX, suggesting that AMPA receptor activation is required for the antidepressant-like actions of (2R,6R)-HNK, similar to the findings with ketamine. They also demonstrated that application of (2R,6R)-HNK increased AMPA receptor-mediated excitatory postsynaptic potentials in hippocampal CA1 slices and upregulated the expression of GluA1 and GluA2 AMPA receptor subunits in hippocampal synaptoneurosomes (Zanos et al. 2016), supporting the involvement of the activation of AMPA receptor in the antidepressant-like actions of (2R,6R)-HNK. A recent study also reported that time- and concentration-dependent increase in GluA1 expression induced by (2R,6R)-HNK (≥0.1 μM at ≥90 min) in cultured cortical pyramidal neurons in rats, in which the (2R,6R)-HNK's concentrations that increased GluA1 expression are consistent with its maximal unbound brain (2R,6R)-HNK concentrations (0.92–4.84 μM) at reportedly efficacious doses of ketamine or (2R,6R)-HNK in animal models of depression in mice (Zanos et al. 2016; Fukumoto et al. 2019; Shaffer et al. 2019). Importantly, unlike ketamine, (2R,6R)-HNK does not have affinity for NMDA receptor at effective dose in rodents (Ki >10 μM for NMDA receptor) (Zanos et al. 2016; Suzuki et al. 2017; Lumsden et al. 2019; Moaddel et al. 2013), which may be the reason that (2R,6R)-HNK does not induce ketamine-like side effects. Therefore, these findings suggest that the antidepressant-like actions of (2R,6R)-HNK do not depend on inhibition of NMDA receptor, primary target of ketamine, but involve activations of AMPA receptor (Zanos et al. 2016). In addition, a recent study reported that (2R,6R)-HNK (even at 10 μM) did not bind orthosterically to or have functional activity at AMPA receptor (Hare et al. 2020), hinting that (2R,6R)-HNK activates AMPA receptor indirectly, not directly, and leading to induction of its antidepressant-like actions.

Brain-derived neurotrophic factor (BDNF), a downstream target of AMPA receptor, is necessary for the antidepressant-like effects of rapid and sustained antidepressant agents including ketamine (Autry et al. 2011; Liu et al. 2017). Recently, we showed that antidepressant-like effects of (2R,6R)-HNK were blocked both in BDNF Val66Met knock-in mice, in which the processing and activity-dependent release of BDNF is blocked (Chen et al. 2006), and by infusion of a BDNF neutralizing antibody into mPFC (Fukumoto et al. 2019). Moreover, central administration of BDNF produces sustained antidepressant-like effects in rodents (Shirayama et al. 2002; Hoshaw et al. 2005; Naumenko et al. 2012). These findings indicate that BDNF release in mPFC is required for (2R,6R)-HNK to exert antidepressant-like effects. This is supported by our and previous studies of primary cortical neurons and hippocampal tissue slices (Zanos et al. 2016; Fukumoto et al. 2019). In addition, (2R,6R)-HNK is reported to increase the expression of mature BDNF in hippocampal synaptoneurosomes (Zanos et al. 2016). These reports provide direct evidences that (2R,6R)-HNK increases the release and the expression of BDNF. Our study also reported that BDNF release induced by (2R,6R)-HNK was blocked by treatment of the L-type voltage-dependent Ca^{2+} channels (VDCCs) antagonist verapamil, which is consistent with evidence that activity-dependent BDNF release requires VDCC activation (Fukumoto et al. 2019; Jourdi et al. 2009). Moreover, we showed that antidepressant-like effects of (2R,6R)-HNK were blocked by VDCC antagonist (Fukumoto et al. 2019). Microinjection of a selective inhibitor of tropomyosin-related kinase B (TrkB) receptor, a high-affinity tyrosine kinase receptor in the downstream of BDNF, ANA 12 into the mPFC is also reported to block the antidepressant-like actions of (2R,6R)-HNK (Fukumoto et al. 2019). Therefore, those findings suggest that the activation of BDNF/TrkB receptor signaling in mPFC through the AMPA receptor and VDCC is essential for (2R,6R)-HNK to induce the antidepressant-like actions.

Previous studies demonstrated that ketamine stimulated mechanistic target of rapamycin complex1 (mTORC1) signaling in the mPFC through the activity-dependent release of BDNF and TrkB receptor signaling (Yang et al. 2017; Li et al. n.d.; Lepack et al. 2016; Ashley et al. 2014; Koike et al. 2013; Harraz et al. 2016; Miller et al. 2014; Jiang et al. 2018). A recent study showed that (2R,6R)-HNK increased the phosphorylated and activated form of mTOR in mPFC (Fukumoto et al. 2019). In contrast, Zanos et al. (2016) reported that (2R,6R)-HNK had no effect on mTORC1 signaling. This discrepancy could be caused by the differences in the brain region (hippocampus vs. mPFC) and the time point (60 min vs. 30 min). (2R,6R)-HNK is also reported to increase the levels of phosphorylated p70S6K, a downstream target of mTORC1, through the TrkB receptor activation. In addition, the antidepressant-like behavioral actions of (2R,6R)-HNK are blocked by microinjection of the selective mTORC1 inhibitor rapamycin into the mPFC, suggesting the involvement of mTORC1 signaling in the antidepressant-like effects of (2R,6R)-HNK (Fukumoto et al. 2019).

(2R,6R)-HNK has been reported to exert sustained antidepressant-like effects up to 3 to 5 days after the treatment (Zanos et al. 2016; Fukumoto et al. 2019), even though (2R,6R)-HNK is rapidly eliminated from the brain within hours (Zanos et al. 2016). These findings suggest that the long-lasting antidepressant-like effects of (2R,6R)-HNK require to be initiated by the stimulation of signaling pathways that result in rapid and sustained synaptic and behavioral actions. Previous studies reported that the increase levels of synaptic proteins as well as synaptic function in layer V pyramidal neurons in the mPFC were involved in the long-lasting antidepressant-like effects of ketamine (Pham et al. 2018; Li et al. n.d.; Liu and Fuchikami 2013). Consistent with this, (2R,6R)-HNK is also reported to increase the function of spine synapses in layer V pyramidal neurons in the mPFC (Fukumoto et al. 2019). This effect includes the increase of frequency and/or amplitude of 5-HT-induced and hypocretin-induced excitatory postsynaptic currents, which act through apically targeted corticocortical and thalamocortical inputs, respectively (Fukumoto et al. 2019; Aghajanian and Marek 1997; Lambe and Aghajanian 2003). These findings imply the involvement of the enhancement of synaptic function in the actions of (2R,6R)-HNK. In addition, there are further findings implying that the actions of (2R,6R)-HNK require the synaptic plasticity. Rac1, a member of the Rho family of GTPase, is required for BDNF/TrkB signaling-induced synaptic plasticity (Hedrick et al. 2016; Lai et al. 2012) and is necessary for the antidepressant-like actions of BDNF infused into the mPFC (Kato et al. 2018), indicating that BDNF/TrkB signaling induces synaptic plasticity through the activation of Rac1 and leads to antidepressant-like effects. We reported that the antidepressant-like effects of (2R,6R)-HNK as well as ketamine were blocked by microinjection of Rac1 inhibitor NSC 23766 into the mPFC (Fukumoto et al. 2019). These findings suggest that the synaptic plasticity through the activation of Rac1 in the mPFC is necessary for (2R,6R)-HNK to provide the antidepressant-like actions, similar to ketamine. Moreover, it is reported that (2R,6R)-HNK enhances structural plasticity in mesencephalic of mice and human induced pluripotent stem cells (iPSC)-derived dopaminergic neurons through the AMPA receptor/BDNF signaling (Cavalleri et al. 2018) and that (2R,6R)-HNK facilitates dendrite outgrowth in human iPSC-derived neurons through AMPA receptor (Collo et al. 2018). Although (2R,6R)-HNK can promote the synaptic plasticity in dopaminergic neurons, the relationships between these actions of (2R,6R)-HNK and its antidepressant-like actions remain unclear.

Recently, the participation of other mechanisms in the antidepressant-like effects of (2R,6R)-HNK is to be reported. Wray et al. (Wray et al. 2019) reported that the translocation of $G\alpha_S$ from lipid rafts may be involved in the antidepressant-like actions of (2R,6R)-HNK. In addition, Zanos et al. (2019b) reported that (2R,6R)-HNK may provide antidepressant-like actions through the metabotropic glutamate 2 receptor signaling-dependent mechanism.

Conclusion

Accumulated evidences imply that (2R,6R)-HNK activates BDNF/TrkB signaling through the indirect activation of AMPA receptor and subsequently stimulates the signaling pathways that enhance synaptic plasticity in mPFC, leading the antidepressant-like responses (Fig. 1). This mechanism, except absence of NMDA receptor blockade, converges with those of antidepressant-like actions of ketamine (Jourdi et al. 2009; Moghaddam et al. 1997; Wohleb et al. 2017). Recently, other mechanisms of the actions of (2R,6R)-HNK are also reported. However, the initial

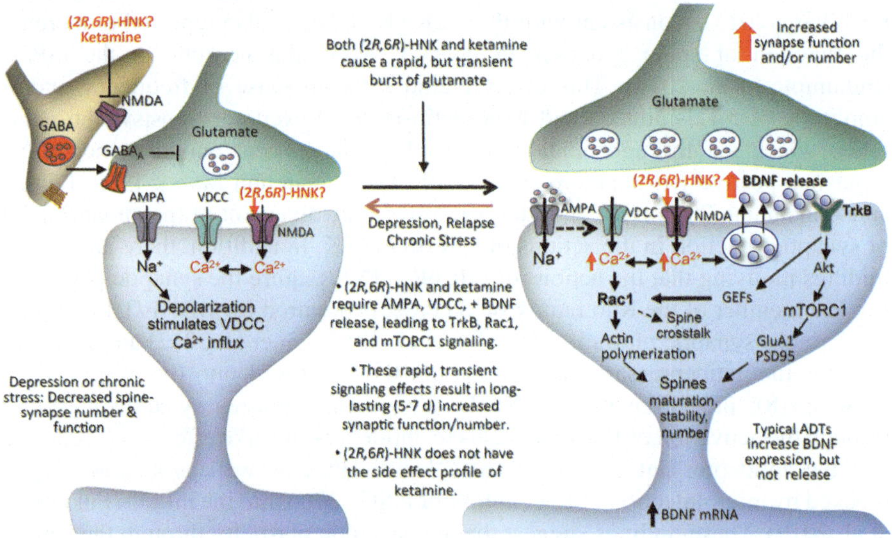

Fig. 1 Model for the cellular mechanisms that underlie the rapid antidepressant-like effects of (2R,6R)-HNK. (2R,6R)-HNK stimulates AMPA receptor and VDCC, causing activity-dependent release of BDNF, activation of TrkB-Rac1, and mTORC1 signaling. This leads to increased synaptic function but not the number of new synapses in layer V pyramidal neurons in the mPFC. These effects are associated with the rapid and long-lasting antidepressant-like effects of (2R,6R)-HNK. The antidepressant-like actions of ketamine are mediated by similar signaling mechanisms, including stimulation of AMPA receptor/VDCC, BDNF release, and subsequent activation of TrkB-Rac1 and mTORC1 signaling, except for the blockade of NMDA receptors on tonic firing GABA interneurons resulting in disinhibition and a transient burst of glutamate

target of (2R,6R)-HNK to induce the antidepressant-like actions is unknown; further researches for its molecular targets and exact mechanism of action are needed and the finding may raise (2R,6R)-HNK as one of the promising candidates for new antidepressant.

This review proposes that (2R,6R)-HNK has the antidepressant-like potential in rodents, but its antidepressant-like profile does not play a main role in the antidepressant actions of ketamine and may be substantially different from those of ketamine. (S)-Ketamine is reported to induce rapid and sustained antidepressant effects in patients with treatment-resistant depression (TRD) (Singh et al. 2016; Daly et al. 2018; Canuso et al. 2018), and US FDA approved (S)-ketamine nasal spray (Spravato) for TRD patients on 5 March 2019 (Turner 2019). Because (2R,6R)-HNK is not generated from (S)-ketamine in humans, this finding also suggests that (2R,6R)-HNK is not necessary for ketamine to exert the antidepressant effects. A recent study demonstrated that (2R,6R)-HNK had favorable oral bioavailability in three species (mouse, rat, and dog) and exhibited p-like efficacy following oral administration in mice (Highland et al. 2019). A clinical trial of (2R,6R)-HNK is currently underway (National Institute of Mental Health, USA). Therefore, whether (2R,6R)-HNK has the therapeutic potential in MDD patients attracts much attention.

Acknowledgments Dr. Duman passed away on February 1, 2020. This article is dedicated to the memory of his mentorship and scientific leadership. K.F. is a former employee of Taisho Pharmaceutical Co., Ltd. and a current employee of Sumitomo Dainippon Pharma.

References

Kessler RC, Berglund P, Demler O, Jin R, Koretz D, Merikangas KR, National Comorbidity Survey Replication et al (2003) The epidemiology of major depressive disorder: results from the National Comorbidity Survey Replication (NCS-R). JAMA 289:3095–3105

Mueller TI, Leon AC, Keller MB, Solomon DA, Endicott J, Coryell W et al (1999) Recurrence after recovery from major depressive disorder during 15 years of observational follow-up. Am J Psychiatry 156:1000–1006

Hasin DS, Sarvet AL, Meyers JL, Saha TD, Ruan WJ, Stohl M et al (2018) Epidemiology of adult DSM-5 major depressive disorder and its specifiers in the United States. JAMA Psychiat 75:336–346

John Rush A, Madhukar H Trivedi, Stephen R Wisniewski, Andrew A Nierenberg, Jonathan W Stewart, Diane Warden, et al. (2006) Acute and longer-term outcomes in depressed outpatients requiring one or several treatment steps: A STAR*D report. Am J Psychiatry 163:1905–1917

Aan Het Rot M, Zarate CA Jr, Charney DS, Mathew SJ (2012) Ketamine for depression: where do we go from here? Biol Psychiatry 72:537–547

Berman RM, Cappiello A, Anand A, Oren DA, Heninger GR, Charney DS et al (2000) Antidepressant effects of ketamine in depressed patients. Biol Psychiatry 47:351–354

Zarate CA Jr, Singh JB, Carlson PJ, Brutsche NE, Ameli R, Luckenbaugh DA et al (2006a) A randomized trial of an N-methyl-D-aspartate antagonist in treatment-resistant major depression. Arch Gen Psychiatry 63:856–864

Ionescu DF, Swee MB, Pavone KJ, Taylor N, Akeju O, Baer L, Nyer M et al (2016) Rapid and sustained reductions in current suicidal ideation following repeated doses of intravenous ketamine: secondary analysis of an open label study. J Clin Psychiatry 77:e719–e725

Krystal JH, Sanacora G, Duman RS (2013) Rapid-acting glutamatergic antidepressants: the path to ketamine and beyond. Biol Psychiatry 73:1133–1141

Sanacora G, Smith MA, Pathak S, Su HL, Boeijinga PH, McCarthy DJ et al (2014) Lanicemine: a low-trapping NMDA channel blocker produces sustained antidepressant efficacy with minimal psychotomimetic adverse effects. Mol Psychiatry 19:978–985

Zarate CA Jr, Singh JB, Quiroz JA, De Jesus G, Denicoff KK, Luckenbaugh DA et al (2006b) A double-blind, placebo-controlled study of memantine in the treatment of major depression. Am J Psychiatry 163:153–155

Hellweg R, Wirth Y, Janetzky W, Hartmann S (2012) Efficacy of memantine in delaying clinical worsening in Alzheimer's disease (AD): responder analyses of nine clinical trials with patients with moderate to severe AD. Int J Geriatr Psychiatry 27:651–656

Zanos P, Moaddel R, Morris PJ, Georgiou P, Fischell J, Elmer GI et al (2016) NMDAR inhibition-independent antidepressant actions of ketamine metabolites. Nature 533:481–486

Hashimoto K (2016) Ketamine's antidepressant action: beyond NMDA receptor inhibition. Expert Opin Ther Targets 20:1389–1392

Suzuki K, Nosyreva E, Hunt KW, Kavalali ET, Monteggia LM (2017) Effects of a ketamine metabolite on synaptic NMDAR function. Nature 546:E1–E3

Yang C, Qu Y, Abe M, Nozawa D, Chaki S, Hashimoto K (2017) (R)-ketamine shows greater potency and longer lasting antidepressant effects than its metabolite (2R,6R)-hydroxynorketamine. Biol Psychiatry 82:e43–e44

Collingridge GL, Lee Y, Bortolotto ZA, Kang H et al (2017) Antidepressant actions of ketamine versus hydroxynorketamine. Biol Psychiatry 81:e65–e67

Zanos P, Moaddel R, Morris PJ, Riggs LM, Highland JN, Georgiou P et al (2018) Ketamine and ketamine metabolite pharmacology: insights into therapeutic mechanisms. Pharmacol. Rev 70:621–630

Zarate CA Jr, Brutsche N, Laje G, Luckenbaugh DA, Vattem Venkata SL, Ramamoorthy A et al (2012) Relationship of ketamine's plasma metabolites with response, diagnosis, and side effects in major depression. Biol Psychiatry 72:331–338

Pham TH, Defaix C, Xu X, Deng S-X, Fabresse N, Alvarez J-C et al (2018) Common neuro-transmission recruited in (R,S)-ketamine and (2R,6R)-hydroxynorketamine-induced sustained antidepressant-like effects. Biol Psychiatry 84:e3–e6

Chou D, Peng H-Y, Lin T-B, Lai C-Y, Hsieh M-C, Wen Y-C et al (2018) (2R,6R)-hydroxynorketamine rescues chronic stress-induced depression-like behavior through its actions in the midbrain periaqueductal gray. Neuropharmacology 139:1–12

Fukumoto K, Manoela V Fogaça, Rong-Jian Liu, Catharine Duman, Taro Kato, Xiao-Yuan Li, et al. (2019) Activity-dependent brain-derived neurotrophic factor signaling is required for the anti-depressant actions of (2R,6R)-hydroxynorketamine. Proc Natl Acad Sci U S A 116:297–302

Shirayama Y, Hashimoto K (2018) Lack of antidepressant effects of (2R,6R)-hydroxynorketamine in a rat learned helplessness model: comparison with (R)-ketamine. Int J Neuropsychopharmacol 21:84–88

Zhang K, Fujita Y, Hashimoto K (2018a) Lack of metabolism in (R)-ketamine's antidepressant actions in a chronic social defeat stress model. Sci Rep 8:4007

Zhang K, Toki H, Fujita Y, Ma M, Chang L, Qu Y et al (2018b) Lack of deuterium isotope effects in the antidepressant effects of (R)-ketamine in a chronic social defeat stress model. Psychopharmacology 235:3177–3185

Yokoyama R, Higuchi M, Tanabe W, Tsukada S, Naito M, Yamaguchi T et al (2020) (S)-norketamine and (2S,6S)-hydroxynorketamine exert potent antidepressant-like effects in a chronic corticosterone-induced mouse model of depression. Pharmacol Biochem Behav 191:172876

Shirayama Y, Hashimoto K (2017) Effects of a single bilateral infusion of R-ketamine in the rat brain regions of a learned helplessness model of depression. Eur Arch Psychiatry Clin Neurosci 267:177–182

Yamaguchi JI, Toki H, Qu Y, Yang C, Koike H, Hashimoto K et al (2018) (2R,6R)-Hydroxynorketamine is not essential for the antidepressant actions of (R)-ketamine in mice. Neuropsychopharmacology 43:1900–1907

Fukumoto K, Toki H, Iijima M, Hashihayata T, Yamaguchi J-I, Hashimoto K et al (2017) Antidepressant potential of (R)-ketamine in rodent models: comparison with (S)-ketamine. J Pharmacol Exp Ther 361:9–16

Zanos P, Highland JN, Liu X, Troppoli TA, Georgiou P, Lovett J et al (2019a) (R)-ketamine exerts antidepressant actions partly via conversion to (2R,6R)-hydroxynorketamine, while causing adverse effects at sub-Anaesthetic doses. Br J Pharmacol 176(14):2573–2592

Fuchikami M, Thomas A, Liu R, Eric S Wohleb, Benjamin B Land, Ralph J DiLeone, et al. (2015) Optogenetic stimulation of infralimbic PFC reproduces ketamine's rapid and sustained antidepressant actions. Proc Natl Acad Sci U S A 112:8106–8111

Gerhard DM, Pothula S, Liu RJ, Wu M, Li XY, Girgenti MJ et al (2020) GABA interneurons are the cellular trigger for ketamine's rapid antidepressant actions. J Clin Invest 130:1336–1349

Burgdorf J, Zhang X-l, Nicholson KL, Balster RL, Leander JD, Stanton PK et al (2013) GLYX-13, a NMDA receptor glycine-site functional partial agonist, induces antidepressant-like effects without ketamine-like side effects. Neuropsychopharmacology 38:729–742

Navarria A, Wohleb ES, Voleti B, Ota KT, Dutheil S, Lepack AE et al (2015) Rapid antidepressant actions of scopolamine: role of medial prefrontal cortex and M1-subtype muscarinic acetylcholine receptors. Neurobiol Dis 82:254–261

Holmes A, Wellman CL (2009) Stress-induced prefrontal reorganization and executive dysfunction in rodents. Neurosci Biobehav Rev 33:773–783

Murrough JW (2012) Ketamine as a novel antidepressant: from synapse to behavior. Clin Pharmacol Ther 91:303–309

Price JL, Drevets WC (2010) Neurocircuitry of mood disorders. Neuropsychopharmacology 35:192–216

Hare BD, Pothula S, DiLeone RJ, Duman RS (2020) Ketamine increases vmPFC activity: effects of (R)- and (S)-stereoisomers and (2R,6R)-hydroxynorketamine metabolite. Neuropharmacology 166:107947

Shaffer CL, Dutra JK, Tseng WC, Weber ML, Bogart LJ, Hales K et al (2019) Pharmacological evaluation of clinically relevant concentrations of (2R,6R)-hydroxynorketamine. Neuropharmacology 153:73–81

Lumsden EW, Troppoli TA, Myers SJ, Zanos P, Aracava Y, Kehr J, Lovett J, Kim S, Wang FH, Schmidt S, Jenne CE, Yuan P, Morris PJ, Thomas CJ, Zarate CA Jr, Moaddel R, Traynelis SF, Pereira EFR, Thompson SM, Albuquerque EX, Gould TD (2019) Antidepressant-relevant concentrations of the ketamine metabolite (2R,6R)-hydroxynorketamine do not block NMDA receptor function. Proc Natl Acad Sci U S A 116(11):5160–5169

Moaddel R, Abdrakhmanova G, Kozak J, Jozwiak K, Toll L, Jimenez L et al (2013) Sub-anesthetic concentrations of (R,S)-ketamine metabolites inhibit acetylcholine-evoked currents in alpha7 nicotinic acetylcholine receptors. Eur J Pharmacol 698:228–234

Autry AE, Adachi M, Nosyreva E, Na ES, Los MF, Cheng P-f et al (2011) NMDA receptor blockade at rest triggers rapid Behavioural antidepressant responses. Nature 475:91–95

Liu R-J, Duman C, Kato T, Hare B, Lopresto D, Bang E et al (2017) GLYX-13 produces rapid antidepressant responses with key synaptic and behavioral effects distinct from ketamine. Neuropsychopharmacology 42:1231–1242

Chen Z-Y, Jing D, Bath KG, Ieraci A, Khan T, Siao C-J et al (2006) Genetic variant BDNF (Val66Met) polymorphism alters anxiety related behavior. Science 314:140–143

Shirayama Y, Chen AC-H, Nakagawa S, Russel DS, Duman RS (2002) Brain-derived neurotrophic factor produces antidepressant effects in behavioral models of depression. J Neurosci 22:3251–3261

Hoshaw BA, Malberg JE, Lucki I (2005) Central administration of IGF-I and BDNF leads to long-lasting antidepressant-like effects. Brain Res 1037:204–208

Naumenko VS, Kondaurova EM, Bazovkina DV, Tsybko AS, Tikhonova MA, Kulikov AV et al (2012) Effect of brain-derived neurotrophic factor on behavior and key members of the brain serotonin system in genetically predisposed to behavioral disorders mouse strains. Neuroscience 214:59–67

Jourdi H, Hsu Y-T, Zhou M, Qin Q, Bi X, Baudry M (2009) Positive AMPA receptor modulation rapidly stimulates BDNF release and increases dendritic mRNA translation. J Neurosci 29:868886–868897

Nanxin Li, Boyoung Lee, Rong-Jian Liu, Mounira Banasr, Jason M Dwyer, Masaaki Iwata, et al. mTOR-dependent synapse formation underlies the rapid antidepressant effects of NMDA antagonists

Lepack AE, Bang E, Lee B, Dwyer JM, Duman RS (2016) Fast-acting antidepressants rapidly stimulate ERK signaling and BDNF release in primary neuronal cultures. Neuropharmacology 111:242–252

Ashley E, Lepack MF, Dwyer JM, Banasr M, Duman RS (2014) BDNF release is required for the behavioral actions of ketamine. Int J Neuropsychopharmacol 18:pyu033

Koike H, Fukumoto K, Iijima M, Chaki S (2013) Role of BDNF/TrkB signaling in antidepressant-like effects of a group II metabotropic glutamate receptor antagonist in animal models of depression. Behav Brain Res 238:48–52

Harraz MM, Tyagi R, Cortés P, Snyder SH (2016) Antidepressant action of ketamine via mTOR is mediated by inhibition of nitrergic Rheb degradation. Mol Psychiatry 21:313–319

Miller OH, Yang L, Wang C-C, Hargroder EA, Zhang Y, Delpire E et al (2014) GluN2B-containing NMDA receptors regulate depression-like behavior and are critical for the rapid antidepressant actions of ketamine. elife 3:e03581

Jiang C, Lin W-J, Sadahiro M, Labonté B, Menard C, Pfau ML et al (2018) VGF function in depression and antidepressant efficacy. Mol Psychiatry 23:1632–1642

Liu R-J, Fuchikami M, Dwyer JM, Lepack AE, Duman RS, Aghajanian GK (2013) GSK-3 inhibition potentiates the synaptogenic and antidepressant-like effects of subthreshold doses of ketamine. Neuropsychopharmacology 38:2268–2277

Aghajanian GK, Marek GJ (1997) Serotonin induces excitatory postsynaptic potentials in apical dendrites of neocortical pyramidal cells. Neuropharmacology 36:589–599

Lambe EK, Aghajanian GK (2003) Hypocretin (orexin) induces calcium transients in single spines postsynaptic to identified thalamocortical boutons in prefrontal slice. Neuron 40:139–150

Hedrick NG, Harward SC, Hall CE, Murakoshi H, McNamara JO, Yasuda R (2016) Rho GTPase complementation underlies BDNF-dependent homo- and heterosynaptic plasticity. Nature 538:104–108

Lai W-O, Wong ASL, Cheung M-C, Xu P, Liang Z, Lok K-C et al (2012) TrkB phosphorylation by Cdk5 is required for activity-dependent structural plasticity and spatial memory. Nat Neurosci 15:1506–1515

Kato T, Fogaça MV, Deyama S, Li X-Y, Fukumoto K, Duman RS (2018) BDNF release and signaling are required for the antidepressant actions of GLYX-13. 2018. Mol Psychiatry 23:2007–2017

Cavalleri L, Merlo Pich E, Millan MJ, Chiamulera C, Kunath T, Spano PF et al (2018) Ketamine enhances structural plasticity in mouse mesencephalic and human iPSC-derived dopaminergic neurons via AMPAR-driven BDNF and mTOR signaling. Mol Psychiatry 23:812–823

Collo G, Cavalleri L, Chiamulera C, Pich EM (2018) (2R,6R)- Hydroxynorketamine promotes dendrite outgrowth in human inducible pluripotent stem cell-derived neurons through AMPA receptor with timing and exposure compatible with ketamine infusion pharmacokinetics in humans. Neuroreport 29:1425–1430

Wray NH, Schappi JM, Singh H, Senese NB, Rasenick MM (2019) NMDAR-independent, cAMP-dependent antidepressant actions of ketamine. Mol Psychiatry 24:1833–1843

Zanos P, Highland JN, Stewart BW, Georgiou P, Jenne CE, Lovett J et al (2019b) (2R,6R)-hydroxynorketamine exerts mGlu 2 receptor-dependent antidepressant actions. Proc Natl Acad Sci U S A 116:6441–6450

Moghaddam B, Adams B, Verma A, Daly D (1997) Activation of glutamatergic neurotransmission by ketamine: A novel step in the pathway from NMDA receptor blockade to dopaminergic and cognitive disruptions associated with the prefrontal cortex. J Neurosci 17:2921–2927

Wohleb ES, Gerhard D, Thomas A, Duman RS (2017) Molecular and cellular mechanisms of rapid-acting antidepressants ketamine and scopolamine. Curr Neuropharmacol 15:11–20

Singh JB, Fedgchin M, Daly E, Xi L, Melman C, De Bruecker G et al (2016) Intravenous esketamine in adult treatment-resistant depression: A double-blind, double-randomization, placebo-controlled study. Biol Psychiatry 80:424–431

Daly EJ, Singh JB, Fedgchin M, Cooper K, Lim P, Shelton RC et al (2018) Efficacy and safety of intranasal Esketamine adjunctive to Oral antidepressant therapy in treatment-resistant depression: A randomized clinical trial. JAMA Psychiat 75:139–148

Canuso CM, Singh JB, Fedgchin M, Alphs L, Lane R, Lim P et al (2018) Efficacy and safety of intranasal Esketamine for the rapid reduction of symptoms of depression and suicidality in patients at imminent risk for suicide: results of a double-blind, randomized, placebo-controlled study. Am J Psychiatry 175:620–630

Turner EH (2019) Esketamine for treatment-resistant depression: seven concerns about efficacy and FDA approval. Lancet Psychiatry:6977–6979

Highland JN, Morris PJ, Zanos P, Lovett J, Ghosh S, Wang AQ et al (2019) Mouse, rat, and dog bioavailability and mouse Oral antidepressant efficacy of (2R,6R)-hydroxynorketamine. J Psychopharmacol 33:12–24

Moghaddam B, Adams B, Verma A, Daly D (1997) Activation of glutamatergic neurotransmission by ketamine: A novel step in the pathway from NMDA receptor blockade to dopaminergic and cognitive disruptions associated with the prefrontal cortex. J Neurosci 17(8):2921–2927
Wohleb ES, Gerhard D, Thomas A, Duman RS (2017) Molecular and cellular mechanisms of rapid-acting antidepressants ketamine and scopolamine. Curr Neuropharmacol 15(1):11–20
[illegible]
[illegible]
[illegible]
[illegible]

Predictable Biomarkers for Rapid-Acting Antidepressant Response to Ketamine

Yunfei Tan and Kenji Hashimoto

Abstract The *N*-methyl-D-aspartate receptor (NMDAR) antagonist (*R,S*)-ketamine elicits rapid-acting and sustained antidepressant actions for treatment-resistant patients with major depressive disorder (MDD) and bipolar disorder (BD) although it is known to cause transient adverse effects such as psychotomimetic and dissociative effects in humans. Identifying specific biomarkers for MDD and BD could help establish accurate diagnoses that form the basis for more appropriate treatment. In this chapter, we discuss the potential biomarkers that can predict the rapid antidepressant response to (*R,S*)-ketamine.

Keywords BDNF, Inflammation · Body mass index · Kynurenine · Gamma power · Structural and functional connectivity · Sleep · Cognition

Introduction

Mental disorders account for 22.8% of the global disease burden (DALYs et al. 2015). The leading cause of this disability is depression (Friedrich 2017), which is the most prevalent mood disorder characterized by a persistent feeling of sadness or lack of interest in external stimuli. Major depressive disorder (MDD) affects millions of individuals worldwide. The economic burden of MDD in the USA has been

Y. Tan
Division of Clinical Neuroscience, Chiba University Center for Forensic Mental Health, Chiba, Japan

Department of Psychiatry, Zhejiang Provincial People's Hospital, People's Hospital of Hangzhou Medical College, Hangzhou, China

K. Hashimoto (✉)
Division of Clinical Neuroscience, Chiba University Center for Forensic Mental Health, Chiba, Japan
e-mail: hashimoto@faculty.chiba-u.jp

K. Hashimoto, M. Manto (eds.), *New Rapid-acting Antidepressants*, Contemporary Clinical Neuroscience,
https://doi.org/10.1007/978-3-030-79790-4_3

estimated to be more than US $210.5 billion, with approximately 45%, 5%, and 50% attributable to direct, suicide-related, and workplace costs, respectively (Greenberg et al. 2015).

Although pharmacological and nonpharmacological treatments are currently available, antidepressants have been used more frequently than psychological interventions. Several antidepressants such as selective serotonin reuptake inhibitors, serotonin norepinephrine reuptake inhibitors, and noradrenergic and specific serotonergic antidepressants are available for treatment. However, approximately one-third of patients fail to respond. In addition, most currently available antidepressants may take several weeks or months to improve depressive symptoms in MDD patients.

More than 40% of patients with MDD would not respond to appropriate antidepressant therapy for an adequate duration. Further, approximately 50% of patients with MDD fail to achieve sustained remission despite various medication trials (Thomas et al. 2015). At present, the delayed onset of antidepressants has been the main limitation in the treatment of depression. MDD patients and physicians may prematurely discontinue a medication that is not perceived as effective. Delayed onset, low remission rate, high recurrence rate, and potential long-term consequences complicate treatment and emphasize the need for new treatment options (Schoeman et al. 2017). These benefits of rapid-acting antidepressant effect include alleviating patient suffering, reducing hospital stay length, reducing the risk of drug-drug interactions and multitargeted drug combinations, and, most importantly, preserving life by decreasing suicide risk.

Biomarkers are frequently used to predict or identify therapeutic responses in the fields of medicine. Identifying specific biomarkers for MDD and BD could help establish accurate diagnoses that form the basis for more appropriate treatment (Hashimoto 2015a, 2014). Moreover, the US Food and Drug Administration (FDA) defines a biomarker as a defining characteristic that can be measured to indicate normal biological processes, pathogenic processes, or responses to an exposure or intervention, including therapeutic interventions.

The *N*-methyl-D-aspartate receptor (NMDAR) antagonist (*R,S*)-ketamine is the first exemplar of a rapid-acting antidepressant for treatment-resistant depression (TRD) for MDD and BD (Hashimoto 2019, 2020; Krystal et al. 2019; Wei et al. 2021; Zhang and Hashimoto 2019). Thus, this chapter briefly summarizes the potential biomarkers that can predict the rapid antidepressant response to (*R,S*)-ketamine (Fig. 1).

Brain-Derived Neurotrophic Factor

Brain-derived neurotrophic factor (BDNF) plays a crucial role in the pathophysiology of MDD and the therapeutic action of antidepressants (Hashimoto 2010; 2015a; 2019; Hashimoto et al. 2004). MDD patients who were Val/Val carriers (rs6265) in the *BDNF* gene exhibited a higher (*R,S*)-ketamine response compared

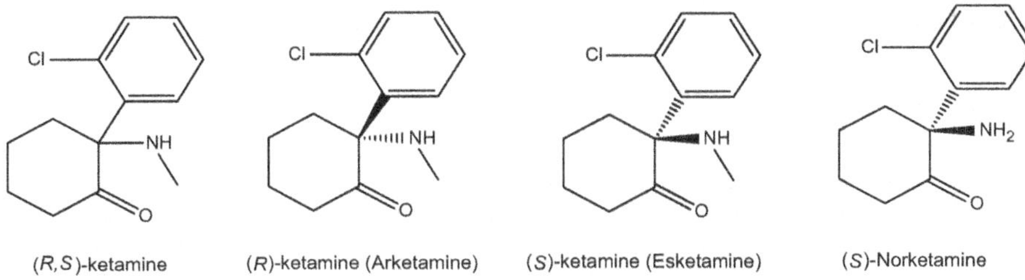

Fig. 1 Chemical structure of (*R*,*S*)-ketamine, (*R*)-ketamine (Arketamine), (*S*)-ketamine (Esketamine), (*S*)-norketamine

with Met carriers (Laje et al. 2012). Due to the ethnic differences in polymorphism distribution (Shimizu et al. 2004), some studies based on the Han population did not replicate this result (Su et al. 2017).

Gururajan et al. (2016) found that microRNAs were unaffected by (*R*,*S*)-ketamine treatment and identified no microRNAs at baseline which can predict therapeutic response to (*R*,*S*)-ketamine. Moreover, Chen et al. (2021a) performed the gene-based genome-wide association study in patients with TRD. They found that specific SNPs and whole genes involved in BDNF and its receptor TrkB (i.e., rs2049048 in *BDNF* gene and rs10217777 in *NTRK2* gene) and the glutamatergic and GABAergic systems (i.e., rs16966731 in GRIN2A gene) were associated with the rapid (within 240 min) antidepressant effect of (*R*,*S*)-ketamine infusion.

Allen et al. (2015) reported that serum BDNF was significantly raised only at 1 week following the first (*R*,*S*)-ketamine infusion in those classified as responders. Interestingly, subsequent ketamine infusions did not elevate serum BDNF levels.

Shank3

Shank3, a postsynaptic density scaffolding and tethering protein, has been implicated in glutamatergic neurotransmission. It is reported that Shank3 may be associated with MDD and BP. Ortiz et al. (2015) reported that plasma levels of Shank3 at baseline predicted antidepressant response to (*R*,*S*)-ketamine. Furthermore, the results showed that plasma levels of Shank3 at baseline were related to the right amygdala volume and glucose metabolism in the hippocampus and amygdala. In the future, carrying out controlled research with a larger sample size is necessary to explore and verify the relationship between Shank3 and the therapeutic response to (*R*,*S*)-ketamine.

Dissociation

(*R,S*)-ketamine has a short-term dissociation effect such as consciousness of individual reporting and change in self-perception and the environment (Metzner and Domino 2010). Moreover, Luckenbaugh et al. (2014) reported that dissociation symptoms measured by the Clinician-Administered Dissociation State Scale (Bremner et al. 1998) were associated with antidepressant improvement 230 min and 7 days after (*R,S*)-ketamine infusion. In contrast, (*R,S*)-ketamine's antidepressant effects were not mediated by the dissociative depersonalization subtype in TRD patients (Acevedo-Diaz et al. 2020). Furthermore, a systematic review using eight papers revealed that the relationship between dissociation and antidepressant effects of (*R,S*)-ketamine is inconsistent (Mathai et al. 2020). However, a recent review concludes that dissociation is not necessary for antidepressant response to (*R,S*)-ketamine (Ballard and Zarate Jr 2020).

(*S*)-ketamine (esketamine; Fig. 1) is a more potent enantiomer of (*R,S*)-ketamine at NMDAR. In 2019, Johnson & Johnson's esketamine nasal spray received the US and UK FDA approval as an adjunctive treatment for TRD. Esketamine is well-known to cause dissociation in TRD patients (Daly et al. 2018; Singh et al. 2016). In contrast, (*R*)-ketamine (arketamine; Fig. 1) has greater and longer-lasting antidepressant-like effects than esketamine in several rodent models of depression (Fukumoto et al. 2017; Yang et al. 2018b, 2015a; Zhang et al. 2014, 2020). Importantly, the side effects of arketamine are less than (*R,S*)-ketamine and esketamine (Hashimoto 2016, 2019, 2020; Wei et al. 2020; Wei et al. 2021; Yang et al. 2019b). A recent pilot study showed that arketamine-induced dissociation was lower in TRD patients after single injection although it caused rapid-acting antidepressant actions (Leal et al. 2021). Consequently, esketamine caused psychotomimetic effects in healthy subjects, whereas arketamine did not cause these side effects in the same subjects (Vollenweider et al. 1997). Thus, it is unlikely that ketamine-induced dissociation is necessary for antidepressant action in TRD patients although further study is needed. Furthermore, conducting a randomized, double-blind study to compare the different effects of arketamine and esketamine in TRD patients is necessary to confirm the role of dissociation in the antidepressant effect of (*R,S*)-ketamine (Hashimoto 2021).

Inflammatory Markers, D-Serine, and Vitamin B12

The identification of blood biomarkers for antidepressant response to (*R,S*)-ketamine is highly useful to avoid unfavorable side effects (i.e., psychotomimetic and dissociative effects and abuse liability) of (*R,S*)-ketamine (Hashimoto 2015b). Consequently, many patients with mood disorders suffer from metabolic syndrome.

Studies show that 38% of TRD patients have metabolic syndrome and revealed that one-third of patients with depression had increased inflammatory markers

(Haroon and Miller 2017). C-reactive protein (CRP) is one of the most important and sensitive markers of the acute-phase response. TRD patients with abnormal levels of plasma CRP had a threefold increased risk of metabolic syndrome (Godin et al. 2019). A systematic review of predictors of antidepressant response in TRD patients suggested that the inflammatory markers such as interleukin (IL)-6, CRP, and high-sensitivity CRP (hsCRP) may predict response to antidepressants that have anti-inflammatory properties.

Yang et al. (2019a) reviewed CRP, hsCRP, and IL-6, which are all inflammatory markers. The authors concluded that these inflammatory markers can predict the response to antidepressants with anti-inflammatory properties. Several animal studies suggested that (*R,S*)-ketamine may have anti-inflammatory effects (Mastrodonato et al. 2020; Xie et al. 2017; Xiong et al. 2019; Zhang et al. 2018). Moreover, Yang et al. (2015b) reported that serum IL-6 is a biomarker with the ability to predict the effect of (*R,S*)-ketamine in patients with TRD. In contrast, other groups have reported that changes in cytokine levels are not correlated with the antidepressant response of (*R,S*)-ketamine. However, fibroblast growth factor changes were related to antidepressant response (Park et al. 2017; Rong et al. 2018). Cytokines were not related to treatment response although they changed transiently in the blood level after (*R,S*)-ketamine infusion (Kiraly et al. 2017). Another study reported that a decrease in tumor necrosis factor (TNF)-α between baseline and 40 min postinfusion of (*R,S*)-ketamine was positively correlated with depression relief across time in the (*R,S*)-ketamine-treated group, suggesting that alteration in inflammatory cytokines may play a direct role (Chen et al. 2018). Recently, Zhan et al. observed the changes of anti-inflammatory factors including granulocyte-macrophage colony-stimulating factor, fractalkine, interferon-gamma, IL-10, IL-12p70, IL-17A, IL-1 β, IL-2, IL-4, IL-23, IL-5, IL-6, IL-7, and TNF-α after repeated injection of (*R,S*)-ketamine. Consequently, these anti-inflammatory factors decreased (all $p < 0.05$). Among them, the changes of IL-17A ($r = -0.259$, $p = 0.046$) and IL-6 ($r = -0.262$, $p = 0.043$) were related to the improvement of depressive symptoms (Zhan et al. 2020). Although inflammatory biomarkers for ketamine's responder are currently inconsistent, further study using a large sample size is needed.

Kadriu et al. (2018) measured the levels of peripheral bone markers in TRD patients and healthy controls at four points (baseline, 230 min, day 1, and day 3) after (*R,S*)-ketamine infusion. They found that (*R,S*)-ketamine infusion substantially raised both the ratio of osteoprotegerin/receptor activator of nuclear factor kappa-B ligand (RANKL) and the levels of plasma osteopontin and substantially reduced RANKL levels. However, the bone marker levels in healthy controls remained unchanged.

D-serine is a potent endogenous co-agonist at the NMDAR (Hashimoto 2014). Moreover, Hashimoto et al. (2016) found that the serum levels of both serine enantiomers in the serum of patients with MDD were raised compared with controls. In 2015, Moaddel et al. (2015) found that baseline D-serine plasma concentrations were significantly lower in (*R,S*)-ketamine responders compared with nonresponders.

Permoda-Osip et al. (2013) reported that the vitamin B12 level was significantly higher in responders compared with the nonresponse patients with bipolar depression after (*R,S*)-ketamine infusion. In contrast, another study did not replicate these results (Lundin et al. 2014). Thus, further study using a larger sample size is needed.

Body Mass Index

The relationship between obesity, depression, and treatment response is suggested. However, the relationship between depression, inflammation, metabolism, and ketamine's action remains unclear. Increased body mass index (BMI) was assumed to be a predictive marker of antidepressant response to (*R,S*)-ketamine (Niciu et al. 2014). In contrast, the presence of metabolic syndrome was negatively correlated with (*R,S*)-ketamine response (Dale et al. 2020). A post hoc analysis from a multisite, randomized, double-blind, and placebo-controlled trial showed that higher BMI and obesity in TRD patients were inversely associated with a robust acute antidepressant response to (*R,S*)-ketamine (Freeman et al. 2020). Consequently, further study is needed to confirm that BMI is a predictor for antidepressant response to (*R,S*)-ketamine or its enantiomers.

Kynurenine Pathway

An increasing body of evidence suggests that abnormal levels of kynurenine (KYN) pathway metabolites are related to depression. KYN is metabolized into kynurenic acid (KYNA), an NMDAR blocker. The KYNA levels and KYNA/KYN ratio at 24 h after (*R,S*)-ketamine infusions were distinctly raised, which related to the relief of depressive symptoms at 24 h, 13 days, and 26 days. They concluded that early alterations of the serum KYNA level and the KYNA/KYN ratio could be potential predictable markers of rapid antidepressant effect of (*R,S*)-ketamine (Zhou et al. 2018a). Nonetheless, performing a larger sample clinical trial is necessary to investigate the role of the kynurenine pathway in predicting the rapid-onset (*R,S*)-ketamine antidepressant.

Gamma Power

Gamma power is a marker of GABA system inhibition and excitatory glutamatergic neurotransmission. It increased after (*R,S*)-ketamine infusion (Gandal et al. 2012; Gilbert and Zarate Jr 2020; Kadriu et al. 2020). Cornwell et al. (2012) reported that (*R,S*)-ketamine responders in TRD patients showed a uniform increase in gamma power post-ketamine infusion compared to baseline. Subsequently, Nugent et al.

(2019b) reported that delayed measurement of gamma power would be a biomarker for (*R,S*)-ketamine responder. Interestingly, TRD patients with lower baseline gamma power levels had a better antidepressant response to (*R,S*)-ketamine compared with higher baseline gamma levels (Nugent et al. 2019a). Moreover, widespread (*R,S*)-ketamine-induced reductions in connectivity in the alpha and beta bands did not correlate with the magnitude of antidepressant response (Nugent et al. 2020). In contrast, a study showed that the variance explained by high gamma power was significantly related to (*R,S*)-ketamine nonresponse, suggesting that the decrease in subgenual anterior cingulate cortex (ACC) BOLD signal is potentially noise and unrelated to (*R,S*)-ketamine's antidepressant effect (McMillan et al. 2020).

A preclinical study demonstrated that (*S*)-norketamine (Fig. 1), a major metabolite from esketamine, showed esketamine-like antidepressant effects in several rodent models of depression and did not cause baseline gamma-band oscillation increase of esketamine in mouse brain (Hashimoto and Yang 2019; Yang et al. 2018a). A recent study using large-scale imaging of neural activity revealed that (*R,S*)-ketamine-induced dissociation may be associated with a 1- to 3-Hz rhythm in layer 5 neurons of the retrosplenial cortex in mouse brain (Vesuna et al. 2020). Thus, further study is needed to confirm the baseline cortical rhythm as a predictable marker for antidepressant response to (*R,S*)-ketamine as well as its two enantiomers in the human brain including the retrosplenial cortex (Hashimoto 2021).

Neuroimaging

Increasing pieces of literature exist on neuroimaging biomarkers of (*R,S*)-ketamine's antidepressant response. (*R,S*)-ketamine preferentially affects the prefrontal cortex (PFC), hippocampus, and ACC of the human brain (Abdallah et al. 2018; Ionescu et al. 2018; Salvadore et al. 2009; Zhong et al. 2011).

Abdallah et al. (2015) reported a correlation between uncorrected left hippocampal volumes and the antidepressant efficacy of (*R,S*)-ketamine in MDD patients ($n = 13$) 24 h after infusion. In contrast, Niciu et al. (2017) reported that the raw and corrected subcortical volume did not associate with an antidepressant response at any time after (*R,S*)-ketamine infusion. A recent study using six infusions demonstrated that larger volumes in the right thalamus and left subiculum head of the hippocampus could predict antidepressant response to (*R,S*)-ketamine in MDD patients (Zhou et al. 2020a). Furthermore, the same group reported no association between alteration in inflammatory markers and changes in hippocampal volume from baseline to day 13 after six (*R,S*)-ketamine infusions (Zhou et al. 2020b). Moreover, Herrera-Melendez et al. (2021) investigated the relationship between baseline volume of the bilateral rostral anterior cingulate cortex (rACC) and antidepressant response to (*R,S*)-ketamine by the whole-brain regression analysis. It was found that a greater baseline volume of the bilateral rACC significantly predicts rapid antidepressant.

Structural and Functional Connectivity

Changes in neural circuits, especially in the pathways connecting the substrate foramen, inner frontal brain, edge steroids, and abdominal synapses, are widely related to the MDD pathophysiology (Phillips et al. 2015; Price and Drevets 2012). The neuroimaging features of white matter (WM) tract microstructure may relate to therapeutic response and treatment prognosis. Furthermore, Vasavada et al. examined regional differences in WM microstructure in advance (*R,S*)-ketamine infusion and found significant differences in fractional anisotropy and radial and mean diffusivities in nonresponders compared with responders and controls (Vasavada et al. 2016).

Local synaptic changes induced by (*R,S*)-ketamine in the brain have been suggested to lead to large-scale network reconstruction, which may be related to its antidepressant properties (Lv et al. 2016). Moreover, low baseline functional connectivity (FC) between the right PFC and the subgenual anterior cingulate gyrus (sgACC) predicts (*R,S*)-ketamine's antidepressant effects (Gärtner et al. 2019). Thus, this study suggests that the PFC-sgACC connectivity may be a biomarker for (*R,S*)-ketamine's responder although further study using a larger sample size is required.

(*R,S*)-ketamine is reported to increase the FC of frontostriatal in TRD patients toward levels observed in healthy controls (Mkrtchian et al. 2020). Furthermore, Sahib et al. (2020a) found that FC destruction in MDD patients trended toward control patterns after (*R,S*)-ketamine injection. Following serial (*R,S*)-ketamine treatment, a significant FC downregulation was shown between the cerebellum and salience network. Moreover, the response showed increased FC between the cerebellum and the striatum before ketamine infusions, whereas nonresponses showed the opposite pattern (Sahib et al. 2020a).

Another study showed that the reduction in depression scores from baseline to 24 h postinfusion of (*R,S*)-ketamine was related to FC upregulation between the right habenula and a cluster in the same side of the frontal pole and the reduction in depression score following (*R,S*)-ketamine treatment was related to FC upregulation between the right Hb and clusters in the right occipital pole, right temporal pole, right parahippocampal gyrus, and left lateral occipital cortex (Rivas-Grajales et al. 2021).

In 2020, Sahib et al. performed a functional magnetic resonance imaging response inhibition task at baseline and 24 h post-ketamine infusions. The patients who obtained a later remission had lower functional activation in the supplementary motor area (SMA) before treatment, which was distinguished from non-remitters. Moreover, SMA alterations were associated with depressive symptom decline, and early changes in the SMA network may serve as a treatment response biomarker (Sahib et al. 2020b).

Collectively, it is possible that (*R,S*)-ketamine leads to neurofunctional plasticity between distinct neural networks that are disrupted in MDD patients. Therefore, performing further clinical trials is necessary to investigate the effect of ketamine enantiomers on FC in multiple brain regions of MDD patients.

Sleep

(*R,S*)-ketamine affects the increased total sleep as well as slow-wave sleep/slow-wave activity (SWS/SWA; (Duncan Jr and Zarate Jr 2013)(Fig. 2). The SWS/SWA improvement, particularly in the early nighttime, is assumed to be the crucial factor in the rapid antidepressant action of (*R,S*)-ketamine in MDD patients. This SWA increase was closely associated with preclinical synaptic plasticity, increased plasma BDNF (Huber et al. 2007), and (*R,S*)-ketamine's response to MDD patients (Duncan et al. 2013; Huber et al. 2007). The improvement of overall sleep, especially the decrease of the objective electroencephalogram, may be one of the mechanisms of (*R,S*)-ketamine's anti-suicide effect (Vande Voort et al. 2017). Moreover, BDNF is a widely known potential marker of antidepressant response (Hashimoto 2010; Hashimoto et al. 2004; Zhang et al. 2016), indicating that BDNF increase may predict the antidepressant response to (*R,S*)-ketamine.

Furthermore, (*R,S*)-ketamine significantly affects the circadian rhythm system, which may be partly due to its glutamate effect (Fig. 2). The synchronization of light/dark and the internal clock was mediated by glutamate in the retina-thalamic tract. Moreover, small-scale studies showed that patients who responded to (*R,S*)-ketamine showed more advanced phase and stronger amplitude increase during the 24-h exercise, suggesting a more robust circadian rhythm (Duncan Jr et al. 2017; Duncan Jr et al. 2018). Furthermore, higher baseline amplitude and delayed 24-h exercise patterns were related to unresponsiveness. Similarly, low amplitude and slow 24-h activity patterns were also associated with rapid recurrence and transient response to (*R,S*)-ketamine (Duncan Jr et al. 2018). A recent study showed that early

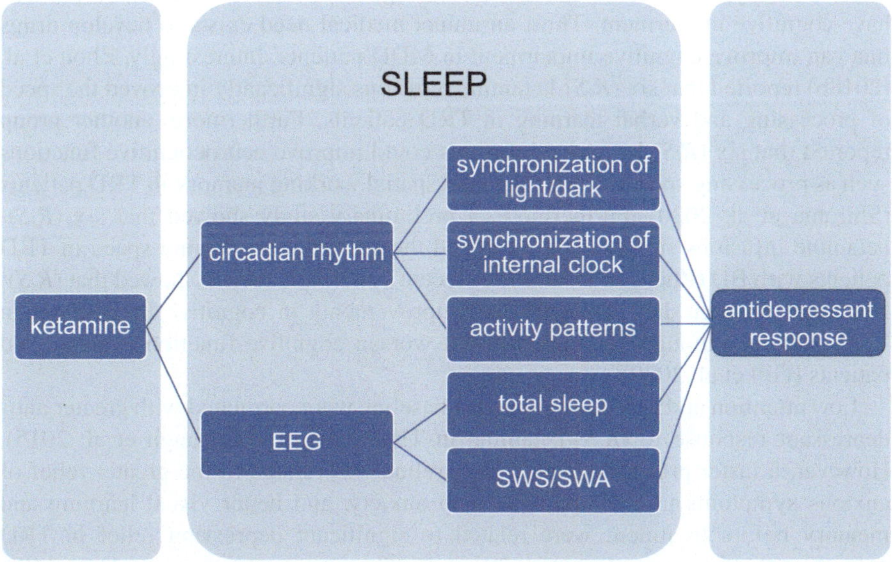

Fig. 2 Sleep as a potential predictable biomarker for antidepressant response to (*R,S*)-ketamine

sleep disturbance improvement by (*R*,*S*)-ketamine may predict the antidepressant action of repeated (R,S)-ketamine infusions and that sleep response was positively associated with serum BDNF levels (Wang et al. 2021).

Sleep deprivation and light therapy in MDD patients may be responsible for the rapid antidepressant effects of (*R*,*S*)-ketamine. Moreover, circadian rhythm disorder is an important MDD biological feature. It often returns to normal with symptom remission (Hasler et al. 2010). Furthermore, patients with abnormal circadian rhythm have been speculated to belong to a subgroup of mood disorders. These patients may be relatively sensitive to (*R*,*S*)-ketamine or other antidepressants that have a rapid impact on the biological clock. Thus, data on the effects of (*R*,*S*)-ketamine on sleep are insufficient and usually limited to short-term intravenous infusion. However, (*R*,*S*)-ketamine treatment may raise the incidence of somnolence compared with the placebo, inconsistent with the conclusion of other studies (Canuso et al. 2018).

Collectively, a few available studies showed that the recovery of structural and functional sleep abnormalities and stability of circadian rhythm may be used as predictors of antidepressant response to (*R*,*S*)-ketamine. Thus, further study is needed to ascertain circadian rhythm as a predictable marker for antidepressant response to (*R*,*S*)-ketamine and its two enantiomers.

Cognition

Cognition is important in the recovery and functional outcome of patients with MDD (McIntyre and Lee 2016). Moreover, approximately 40% of MDD patients have cognitive impairment. Thus, an unmet medical need exists to develop drugs that can improve cognitive impairment in MDD patients. Interestingly, Zhou et al. (2018b) reported that six (*R*,*S*)-ketamine infusions significantly improved the speed of processing and verbal learning in TRD patients. Furthermore, another group reported that six (*R*,*S*)-ketamine infusions could improve neurocognitive functions such as processing speed, set shifting, and spatial working memory in TRD patients (Shiroma et al. 2020). Furthermore, a preliminary study showed that six (*R*,*S*)-ketamine infusions significantly improved the scores of processing speed in TRD patients with BD (Zheng et al. 2020). A recent systematic review showed that (*R*,*S*)-ketamine infusion showed significant improvements in cognitive impairment in TRD patients although ketamine did not worsen cognitive function in depressed patients (Gill et al. 2020).

Low attention and processing speed at baseline were correlated with greater antidepressant response to (*R*,*S*)-ketamine in TRD patients (Murrough et al. 2015). However, a faster processing speed at baseline was related to the greater relief of anxious symptoms in TRD patients with anxiety, and better visual learning and memory before treatment were related to significant depression relief in TRD patients without anxiety (Liu et al. 2019). Furthermore, a recent study showed that the improvement in working memory may predict the anti-suicidal ideation response of (*R*,*S*)-ketamine in TRD patients (Chen et al. 2021b).

A single (*R,S*)-ketamine infusion causes cognitive impairments in the healthy controls (Krystal et al. 1994). The effects of (*R,S*)-ketamine on cognition in depressed patients and healthy controls may be different although further studies on the effects of (*R,S*)-ketamine on cognition of healthy control are needed.

Collectively, it seems that (*R,S*)-ketamine improved cognitive impairments in TRD patients although (*R,S*)-ketamine causes cognitive impairment in healthy controls (Crisanti et al. 2020). Moreover, arketamine, but not esketamine, was recently reported to significantly improve phencyclidine-induced cognitive deficits in mice (Tan et al. 2020). Therefore, conducting a clinical trial is interesting to investigate whether arketamine can improve cognitive impairment in patients with MDD or bipolar depression.

Conclusion

As aforementioned, several predictable biomarkers exist for (*R,S*)-ketamine's antidepressant actions in MDD patients including TRD. Thus, identifying biomarkers that can predict the antidepressant response to (*R,S*)-ketamine and esketamine would be useful in the treatment of patients including TRD because (*R,S*)-ketamine and esketamine have detrimental side effects in humans. Moreover, accumulating evidence shows that the novel rapid-acting candidate arketamine does not cause detrimental side effects of (*R,S*)-ketamine in humans. Therefore, arketamine would be a novel antidepressant without the side effects of (*R,S*)-ketamine and esketamine.

Acknowledgments This study was supported by AMED (to K.H., JP20dm0107119).

Conflict of Interest Dr. Hashimoto is the inventor of filed patent applications on "The use of *R*-ketamine in the treatment of psychiatric diseases," "(*S*)-norketamine and salt thereof as pharmaceutical," "*R*-ketamine and derivative thereof as prophylactic or therapeutic agent for neurodegeneration disease or recognition function disorder," "Preventive or therapeutic agent and pharmaceutical composition for inflammatory diseases or bone diseases," and "*R*-ketamine and its derivatives as a preventive or therapeutic agent for a neurodevelopmental disorder" by the Chiba University. Dr. Hashimoto has received the research support from Dainippon Sumitomo, Otsuka, and Taisho. Dr. Tan declares no conflict of interest.

References

Abdallah CG, Salas R, Jackowski A, Baldwin P, Sato JR, Mathew SJ (2015) Hippocampal volume and the rapid antidepressant effect of ketamine. J Psychopharmacol 29(5):591–595

Abdallah CG, Sanacora G, Duman RS, Krystal JH (2018) The neurobiology of depression, ketamine and rapid-acting antidepressants: is it glutamate inhibition or activation? Pharmacol Ther 190:148–158

Acevedo-Diaz EE, Cavanaugh GW, Greenstein D, Kraus C, Kadriu B, Zarate CA, Park LT (2020) Comprehensive assessment of side effects associated with a single dose of ketamine in treatment-resistant depression. J Affect Disord 263:568–575

Allen AP, Naughton M, Dowling J, Walsh A, Ismail F, Shorten G, Scott L, McLoughlin DM, Cryan JF, Dinan TG, Clarke G (2015) Serum BDNF as a peripheral biomarker of treatment-resistant depression and the rapid antidepressant response: a comparison of ketamine and ECT. J Affect Disord 186:306–311

Ballard ED, Zarate CA Jr (2020) The role of dissociation in ketamine's antidepressant effects. Nat Commun 11(1):6431

Bremner JD, Krystal JH, Putnam FW, Southwick SM, Marmar C, Charney DS, Mazure CM (1998) Measurement of dissociative states with the clinician-administered dissociative states scale (CADSS). J Trauma Stress 11(1):125–136

Canuso CM, Singh JB, Fedgchin M, Alphs L, Lane R, Lim P, Pinter C, Hough D, Sanacora G, Manji H, Drevets WC (2018) Efficacy and safety of intranasal Esketamine for the rapid reduction of symptoms of depression and suicidality in patients at imminent risk for suicide: results of a double-blind, randomized, placebo-controlled study. Am J Psychiatry 175(7):620–630

Chen MH, Li CT, Lin WC, Hong CJ, Tu PC, Bai YM, Cheng CM, Su TP (2018) Rapid inflammation modulation and antidepressant efficacy of a low-dose ketamine infusion in treatment-resistant depression: a randomized, double-blind control study. Psychiatry Res 269:207–211

Chen MH, Kao CF, Tsai SJ, Li CT, Lin WC, Hong CJ, Bai YM, Tu PC, Su TP (2021a) Treatment response to low-dose ketamine infusion for treatment-resistant depression: a gene-based genome-wide association study. Genomics 113(2):507–514

Chen X, Wang M, Hu Y, Zhan Y, Zhou Y, Zheng W, Liu W, Wang C, Zhong X, Li H, Lan X, Ning Y, Zhang B (2021b) Working memory associated with anti-suicidal ideation effect of repeated-dose intravenous ketamine in depressed patients. Advance online publication. Eur Arch Psychiatry Clin Neurosci. https://doi.org/10.1007/s00406-020-01221-z

Cornwell BR, Salvadore G, Furey M, Marquardt CA, Brutsche NE, Grillon C, Zarate CA Jr (2012) Synaptic potentiation is critical for rapid antidepressant response to ketamine in treatment-resistant major depression. Biol Psychiatry 72(7):555–561

Crisanti C, Enrico P, Fiorentini A, Delvecchio G, Brambilla P (2020) Neurocognitive impact of ketamine treatment in major depressive disorder: a review on human and animal studies. J Affect Disord 276:1109–1118

Dale RM, Bryant KA, Thompson NR (2020) Metabolic syndrome rather than body mass index is associated with treatment response to ketamine infusions. J Clin Psychopharmacol 40(1):75–79

Daly EJ, Singh JB, Fedgchin M, Cooper K, Lim P, Shelton RC, Thase ME, Winokur A, Van Nueten L, Manji H, Drevets WC (2018) Efficacy and safety of intranasal Esketamine adjunctive to Oral antidepressant therapy in treatment-resistant depression: a randomized clinical trial. JAMA Psychiat 75(2):139–148

DALYs and HALE Collaborators, Murray CJ, Barber RM, Foreman KJ, Abbasoglu Ozgoren A, Abd-Allah F, Abera SF, Aboyans V, Abraham JP, Abubakar I, Abu-Raddad LJ, Abu-Rmeileh NM, Achoki T, Ackerman IN, Ademi Z, Adou AK, Adsuar JC, Afshin A, Agardh EE, Alam SS et al (2015) Global, regional, and national disability-adjusted life years (DALYs) for 306 diseases and injuries and healthy life expectancy (HALE) for 188 countries, 1990-2013: quantifying the epidemiological transition. Lancet 386(10009):2145–2191

Duncan WC Jr, Slonena E, Hejazi NS, Brutsche N, Yu KC, Park L, Ballard ED, Zarate CA Jr (2017) Motor-activity markers of circadian timekeeping are related to Ketamine's rapid antidepressant properties. Biol Psychiatry 82(5):361–369

Duncan WC Jr, Slonena EE, Hejazi NS, Brutsche N, Park LT, Henter ID, Ballard ED, Zarate CA Jr (2018) Are 24-hour motor activity patterns associated with continued rapid response to ketamine? Neuropsychiatr Dis Treat 14:2739–2748

Duncan WC Jr, Zarate CA Jr (2013) Ketamine, sleep, and depression: current status and new questions. Curr Psychiatry Rep 15(9):394

Duncan WC, Sarasso S, Ferrarelli F, Selter J, Riedner BA, Hejazi NS, Yuan P, Brutsche N, Manji HK, Tononi G, Zarate CA (2013) Concomitant BDNF and sleep slow wave changes indicate ketamine-induced plasticity in major depressive disorder. Int J Neuropsychopharmacol 16(2):301–311

Freeman MP, Hock RS, Papakostas GI, Judge H, Cusin C, Mathew SJ, Sanacora G, Iosifescu DV, DeBattista C, Trivedi MH, Fava M (2020) Body mass index as a moderator of treatment response to ketamine for major depressive disorder. J Clin Psychopharmacol 40(3):287–292

Friedrich MJ (2017) Depression is the leading cause of disability around the world. JAMA 317(15):1517

Fukumoto K, Toki H, Iijima M, Hashihayata T, Yamaguchi JI, Hashimoto K, Chaki S (2017) Antidepressant potential of (*R*)-ketamine in rodent models: comparison with (*S*)-ketamine. J Pharmacol Exp Ther 361(1):9–16

Gandal MJ, Sisti J, Klook K, Ortinski PI, Leitman V, Liang Y, Thieu T, Anderson R, Pierce RC, Jonak G, Gur RE, Carlson G, Siegel SJ (2012) $GABA_B$-mediated rescue of altered excitatory-inhibitory balance, gamma synchrony and behavioral deficits following constitutive NMDAR-hypofunction. Transl Psychiatry 2(7):e142

Gärtner M, Aust S, Bajbouj M, Fan Y, Wingenfeld K, Otte C, Heuser-Collier I, Böker H, Hättenschwiler J, Seifritz E, Grimm S, Scheidegger M (2019) Functional connectivity between prefrontal cortex and subgenual cingulate predicts antidepressant effects of ketamine. Eur Neuropsychopharmacol 29(4):501–508

Gilbert JR, Zarate CA Jr (2020) Electrophysiological biomarkers of antidepressant response to ketamine in treatment-resistant depression: gamma power and long-term potentiation. Pharmacol Biochem Behav 189:172856

Gill H, Gill B, Rodrigues NB, Lipsitz O, Rosenblat JD, El-Halabi S, Nasri F, Mansur RB, Lee Y, McIntyre RS (2020) The effects of ketamine on cognition in treatment-resistant depression: a systematic review and priority avenues for future research. Neurosci Biobehav Rev 120:78–85

Godin O, Bennabi D, Yrondi A, Richieri R, D'Amato T, Bellivier F, Bougerol T, Horn M, Camus V, Courtet P, Doumy O, Genty JB, El-Hage W, Haesebaert F, Holtzmann J, Lancon C, Leboyer M, Llorca PM, Maruani J, Molière F, FondaMental Advanced Centers of Expertise in Resistant Depression (FACE-DR) Collaborators (2019) Prevalence of metabolic syndrome and associated factors in a cohort of individuals with treatment-resistant depression: results from the FACE-DR study. J Clin Psychiatry 80(6):19m12755

Greenberg PE, Fournier AA, Sisitsky T, Pike CT, Kessler RC (2015) The economic burden of adults with major depressive disorder in the United States (2005 and 2010). J Clin Psychiatry 76(2):155–162

Gururajan A, Naughton ME, Scott KA, O'Connor RM, Moloney G, Clarke G, Dowling J, Walsh A, Ismail F, Shorten G, Scott L, McLoughlin DM, Cryan JF, Dinan TG (2016) MicroRNAs as biomarkers for major depression: a role for let-7b and let-7c. Transl Psychiatry 6(8):e862

Haroon E, Miller AH (2017) Inflammation effects on brain glutamate in depression: mechanistic considerations and treatment implications. Curr Top Behav Neurosci 31:173–198

Hashimoto K (2010) Brain-derived neurotrophic factor as a biomarker for mood disorders: an historical overview and future directions. Psychiatry Clin Neurosci 64(4):341–357

Hashimoto K (2014) Blood D-serine levels as a predictive biomarker for the rapid antidepressant effects of the NMDA receptor antagonist ketamine. psychopharmacology (Berl) 231(20):4081–2

Hashimoto K (2015a) Brain-derived neurotrophic factor (BDNF) and its precursor proBDNF as diagnostic biomarkers for major depressive disorder and bipolar disorder. Eur Arch Psychiatry Clin Neurosci 265(1):83–84

Hashimoto K (2015b) Inflammatory biomarkers as differential predictors of antidepressant response. Int J Mol Sci 16(4):7796–7801

Hashimoto K (2016) Letter to the editor: *R*-ketamine: a rapid-onset and sustained antidepressant without risk of brain toxicity. Psychol Med 46(11):2449–2451

Hashimoto K (2019) Rapid-acting antidepressant ketamine, its metabolites and other candidates: a historical overview and future perspective. Psychiatry Clin Neurosci 73(10):613–627

Hashimoto K (2020) Molecular mechanisms of the rapid-acting and long-lasting antidepressant actions of (*R*)-ketamine. Biochem Pharmacol 177:113935

Hashimoto K (2021) Neural rhythm in the retrosplenial cortex during ketamine-induced dissociation. Eur Arch Psychiatry Clin Neurosci 271(3):583–585

Hashimoto K, Shimizu E, Iyo M (2004) Critical role of brain-derived neurotrophic factor in mood disorders. Brain research. Brain Res Rev 45(2):104–114

Hashimoto K, Yang C (2019) Is (*S*)-norketamine an alternative antidepressant for esketamine? Eur Arch Psychiatry Clin Neurosci 269(7):867–868

Hashimoto K, Yoshida T, Ishikawa M, Fujita Y, Niitsu T, Nakazato M, Watanabe H, Sasaki T, Shiina A, Hashimoto T, Kanahara N, Hasegawa T, Enohara M, Kimura A, Iyo M (2016) Increased serum levels of serine enantiomers in patients with depression. Acta Neuropsychiatrica 28(3):173–178

Hasler BP, Buysse DJ, Kupfer DJ, Germain A (2010) Phase relationships between core body temperature, melatonin, and sleep are associated with depression severity: further evidence for circadian misalignment in non-seasonal depression. Psychiatry Res 178(1):205–207

Herrera-Melendez A, Stippl A, Aust S, Scheidegger M, Seifritz E, Heuser-Collier I, Otte C, Bajbouj M, Grimm S, Gärtner M (2021) Gray matter volume of rostral anterior cingulate cortex predicts rapid antidepressant response to ketamine. Eur Neuropsychopharmacol 43:63–70

Huber R, Tononi G, Cirelli C (2007) Exploratory behavior, cortical BDNF expression, and sleep homeostasis. Sleep 30(2):129–139

Ionescu DF, Felicione JM, Gosai A, Cusin C, Shin P, Shapero BG, Deckersbach T (2018) Ketamine-associated brain changes: a review of the neuroimaging literature. Harv Rev Psychiatry 26(6):320–339

Kadriu B, Ballard ED, Henter ID, Murata S, Gerlus N, Zarate CA Jr (2020) Neurobiological biomarkers of response to ketamine. Adv Pharmacol 89:195–235

Kadriu B, Gold PW, Luckenbaugh DA, Lener MS, Ballard ED, Niciu MJ, Henter ID, Park LT, De Sousa RT, Yuan P, Machado-Vieira R, Zarate CA (2018) Acute ketamine administration corrects abnormal inflammatory bone markers in major depressive disorder. Mol Psychiatry 23(7):1626–1631

Kiraly DD, Horn SR, Van Dam NT, Costi S, Schwartz J, Kim-Schulze S, Patel M, Hodes GE, Russo SJ, Merad M, Iosifescu DV, Charney DS, Murrough JW (2017) Altered peripheral immune profiles in treatment-resistant depression: response to ketamine and prediction of treatment outcome. Transl Psychiatry 7(3):e1065

Krystal JH, Abdallah CG, Sanacora G, Charney DS, Duman RS (2019) Ketamine: a paradigm shift for depression research and treatment. Neuron 101(5):774–778

Krystal JH, Karper LP, Seibyl JP, Freeman GK, Delaney R, Bremner JD, Heninger GR, Bowers MB Jr, Charney DS (1994) Subanesthetic effects of the noncompetitive NMDA antagonist, ketamine, in humans. Psychotomimetic, perceptual, cognitive, and neuroendocrine responses. Arch Gen Psychiatry 51(3):199–214

Laje G, Lally N, Mathews D, Brutsche N, Chemerinski A, Akula N, Kelmendi B, Simen A, McMahon FJ, Sanacora G, Zarate C Jr (2012) Brain-derived neurotrophic factor Val66Met polymorphism and antidepressant efficacy of ketamine in depressed patients. Biol Psychiatry 72(11):e27–e28

Leal GC, Bandeira ID, Correia-Melo FS, Telles M, Mello RP, Vieira F, Lima CS, Jesus-Nunes AP, Guerreiro-Costa L, Marback RF, Caliman-Fontes AT, Marques B, Bezerra M, Dias-Neto AL, Silva SS, Sampaio AS, Sanacora G, Turecki G, Loo C, Lacerda A, Quarantini LC (2021) Intravenous arketamine for treatment-resistant depression: open-label pilot study. Eur Arch Psychiatry Clin Neurosci 271(3):577–582

Liu W, Zhou Y, Zheng W, Wang C, Zhan Y, Lan X, Zhang B, Li H, Chen L, Ning Y (2019) Repeated intravenous infusions of ketamine: Neurocognition in patients with anxious and nonanxious treatment-resistant depression. J Affect Disord 259:1–6

Luckenbaugh DA, Niciu MJ, Ionescu DF, Nolan NM, Richards EM, Brutsche NE, Guevara S, Zarate CA (2014) Do the dissociative side effects of ketamine mediate its antidepressant effects? J Affect Disord 159:56–61

Lundin NB, Niciu MJ, Luckenbaugh DA, Ionescu DF, Richards EM, Vande Voort JL, Brutsche NE, Machado-Vieira R, Zarate CA Jr (2014) Baseline vitamin B12 and folate levels do not predict improvement in depression after a single infusion of ketamine. Pharmacopsychiatry 47(4–5):141–144

Lv Q, Yang L, Li G, Wang Z, Shen Z, Yu W, Jiang Q, Hou B, Pu J, Hu H, Wang Z (2016) Large-scale persistent network reconfiguration induced by ketamine in anesthetized monkeys: relevance to mood disorders. Biol Psychiatry 79(9):765–775

Mastrodonato A, Cohensedgh O, LaGamma CT, McGowan JC, Hunsberger HC, Denny CA (2020) Prophylactic (*R,S*)-ketamine selectively protects against inflammatory stressors. Behav Brain Res 378:112238

Mathai DS, Meyer MJ, Storch EA, Kosten TR (2020) The relationship between subjective effects induced by a single dose of ketamine and treatment response in patients with major depressive disorder: a systematic review. J Affect Disord 264:123–129

McIntyre RS, Lee Y (2016) Cognition in major depressive disorder: a 'Systemically important functional Index' (SIFI). Curr Opin Psychiatry 29(1):48–55

McMillan R, Sumner R, Forsyth A, Campbell D, Malpas G, Maxwell E, Deng C, Hay J, Ponton R, Sundram F, Muthukumaraswamy S (2020) Simultaneous EEG/fMRI recorded during ketamine infusion in patients with major depressive disorder. Prog Neuro-Psychopharmacol Biol Psychiatry 99:109838

Metzner J, Domino KB (2010) Risks of anesthesia or sedation outside the operating room: the role of the anesthesia care provider. Curr Opin Anaesthesiol 23(4):523–531

Mkrtchian A, Evans JW, Kraus C, Yuan P, Kadriu B, Nugent AC, Roiser JP, Zarate CA Jr (2020) Ketamine modulates fronto-striatal circuitry in depressed and healthy individuals. Molecular Psychiatry, Advance online publication. https://doi.org/10.1038/s41380-020-00878-1

Moaddel R, Luckenbaugh DA, Xie Y, Villaseñor A, Brutsche NE, Machado-Vieira R, Ramamoorthy A, Lorenzo MP, Garcia A, Bernier M, Torjman MC, Barbas C, Zarate CA Jr, Wainer IW (2015) D-serine plasma concentration is a potential biomarker of (*R,S*)-ketamine antidepressant response in subjects with treatment-resistant depression. Psychopharmacology 232(2):399–409

Murrough JW, Burdick KE, Levitch CF, Perez AM, Brallier JW, Chang LC, Foulkes A, Charney DS, Mathew SJ, Iosifescu DV (2015) Neurocognitive effects of ketamine and association with antidepressant response in individuals with treatment-resistant depression: a randomized controlled trial. Neuropsychopharmacology 40(5):1084–1090

Niciu MJ, Iadarola ND, Banerjee D, Luckenbaugh DA, Park M, Lener M, Park L, Ionescu DF, Ballard ED, Brutsche NE, Akula N, McMahon FJ, Machado-Vieira R, Nugent AC, Zarate CA Jr (2017) The antidepressant efficacy of subanesthetic-dose ketamine does not correlate with baseline subcortical volumes in a replication sample with major depressive disorder. J Psychopharmacol 31(12):1570–1577

Niciu MJ, Luckenbaugh DA, Ionescu DF, Guevara S, Machado-Vieira R, Richards EM, Brutsche NE, Nolan NM, Zarate CA Jr (2014) Clinical predictors of ketamine response in treatment-resistant major depression. J Clin Psychiatry 75(5):e417–e423

Nugent AC, Ballard ED, Gilbert JR, Tewarie PK, Brookes MJ, Zarate CA Jr (2020) The effect of ketamine on electrophysiological connectivity in major depressive disorder. Front Psych 11:519

Nugent AC, Ballard ED, Gould TD, Park LT, Moaddel R, Brutsche NE, Zarate CA Jr (2019a) Ketamine has distinct electrophysiological and behavioral effects in depressed and healthy subjects. Mol Psychiatry 24(7):1040–1052

Nugent AC, Wills KE, Gilbert JR, Zarate CA Jr (2019b) Synaptic potentiation and rapid antidepressant response to ketamine in treatment-resistant major depression: a replication study. Psychiatry Res Neuroimaging 283:64–66

Ortiz R, Niciu MJ, Lukkahati N, Saligan LN, Nugent AC, Luckenbaugh DA, Machado-Vieira R, Zarate CA Jr (2015) Shank3 as a potential biomarker of antidepressant response to ketamine and its neural correlates in bipolar depression. J Affect Disord 172:307–311

Park M, Newman LE, Gold PW, Luckenbaugh DA, Yuan P, Machado-Vieira R, Zarate CA Jr (2017) Change in cytokine levels is not associated with rapid antidepressant response to ketamine in treatment-resistant depression. J Psychiatr Res 84:113–118

Permoda-Osip A, Dorszewska J, Bartkowska-Sniatkowska A, Chlopocka-Wozniak M, Rybakowski JK (2013) Vitamin B12 level may be related to the efficacy of single ketamine infusion in bipolar depression. Pharmacopsychiatry 46(6):227–228

Phillips ML, Chase HW, Sheline YI, Etkin A, Almeida JR, Deckersbach T, Trivedi MH (2015) Identifying predictors, moderators, and mediators of antidepressant response in major depressive disorder: neuroimaging approaches. Am J Psychiatry 172(2):124–138

Price JL, Drevets WC (2012) Neural circuits underlying the pathophysiology of mood disorders. Trends Cogn Sci 16(1):61–71

Rivas-Grajales AM, Salas R, Robinson ME, Qi K, Murrough JW, Mathew SJ (2021) Habenula connectivity and intravenous ketamine in treatment-resistant depression. Inter J Neuropsycho pharmacol 24(5):383–391

Rong C, Park C, Rosenblat JD, Subramaniapillai M, Zuckerman H, Fus D, Lee YL, Pan Z, Brietzke E, Mansur RB, Cha DS, Lui L, McIntyre RS (2018) Predictors of response to ketamine in treatment resistant major depressive disorder and bipolar disorder. Int J Environ Res Public Health 15(4):771

Sahib AK, Loureiro JR, Vasavada M, Anderson C, Kubicki A, Wade B, Joshi SH, Woods RP, Congdon E, Espinoza R, Narr KL (2020a) Modulation of the functional connectome in major depressive disorder by ketamine therapy. Psycholog Med Advance online publication:1–10. https://doi.org/10.1017/S0033291720004560

Sahib AK, Loureiro JR, Vasavada MM, Kubicki A, Wade B, Joshi SH, Woods RP, Congdon E, Espinoza R, Narr KL (2020b) Modulation of inhibitory control networks relate to clinical response following ketamine therapy in major depression. Transl Psychiatry 10(1):260

Salvadore G, Cornwell BR, Colon-Rosario V, Coppola R, Grillon C, Zarate CA Jr, Manji HK (2009) Increased anterior cingulate cortical activity in response to fearful faces: a neurophysiological biomarker that predicts rapid antidepressant response to ketamine. Biol Psychiatry 65(4):289–295

Schoeman JC, Steyn SF, Harvey BH, Brink CB (2017) Long-lasting effects of fluoxetine and/or exercise augmentation on bio-behavioural markers of depression in pre-pubertal stress sensitive rats. Behav Brain Res 323:86–99

Shimizu E, Hashimoto K, Iyo M (2004) Ethnic difference of the BDNF 196G/a (val66met) polymorphism frequencies: the possibility to explain ethnic mental traits. American journal of medical genetics. Part B. Neuropsychiatric Genetics 126B(1):122–123

Shiroma PR, Thuras P, Wels J, Albott CS, Erbes C, Tye S, Lim KO (2020) Neurocognitive performance of repeated versus single intravenous subanesthetic ketamine in treatment resistant depression. J Affect Disord 277:470–477

Singh JB, Fedgchin M, Daly E, Xi L, Melman C, De Bruecker G, Tadic A, Sienaert P, Wiegand F, Manji H, Drevets WC, Van Nueten L (2016) Intravenous Esketamine in adult treatment-resistant depression: a double-blind, double-randomization, placebo-controlled study. Biol Psychiatry 80(6):424–431

Su TP, Chen MH, Li CT, Lin WC, Hong CJ, Gueorguieva R, Tu PC, Bai YM, Cheng CM, Krystal JH (2017) Dose-related effects of adjunctive ketamine in Taiwanese patients with treatment-resistant depression. Neuropsychopharmacology 42(13):2482–2492

Tan Y, Fujita Y, Qu Y, Chang L, Pu Y, Wang S, Wang X, Hashimoto K (2020) Phencyclidine-induced cognitive deficits in mice are ameliorated by subsequent repeated intermittent administration of (*R*)-ketamine, but not (*S*)-ketamine: role of BDNF-TrkB signaling. Pharmacol Biochem Behav 188:172839

Thomas SJ, Shin M, McInnis MG, Bostwick JR (2015) Combination therapy with monoamine oxidase inhibitors and other antidepressants or stimulants: strategies for the management of treatment-resistant depression. Pharmacotherapy 35(4):433–449

Vande Voort JL, Ballard ED, Luckenbaugh DA, Bernert RA, Richards EM, Niciu MJ, Park LT, Machado-Vieira R, Duncan WC Jr, Zarate CA Jr (2017) Antisuicidal response following ketamine infusion is associated with decreased nighttime wakefulness in major depressive disorder and bipolar disorder. J Clin Psychiatry 78(8):1068–1074

Vasavada MM, Leaver AM, Espinoza RT, Joshi SH, Njau SN, Woods RP, Narr KL (2016) Structural connectivity and response to ketamine therapy in major depression: a preliminary study. J Affect Disord 190:836–841

Vesuna S, Kauvar IV, Richman E, Gore F, Oskotsky T, Sava-Segal C, Luo L, Malenka RC, Henderson JM, Nuyujukian P, Parvizi J, Deisseroth K (2020) Deep posteromedial cortical rhythm in dissociation. Nature 586(7827):87–94

Vollenweider FX, Leenders KL, Oye I, Hell D, Angst J (1997) Differential psychopathology and patterns of cerebral glucose utilisation produced by (*S*)- and (*R*)-ketamine in healthy volunteers using positron emission tomography (PET). Eur Neuropsychopharmacol J Eur Coll Neuropsychopharmacol 7(1):25–38

Wang M, Zhang B, Zhou Y, Wang C, Zheng W, Liu W, Zhan Y, Lan X, Ning Y (2021) Sleep improvement is associated with the antidepressant efficacy of repeated-dose ketamine and serum BDNF levels: a post-hoc analysis. Pharmacological Reports, PR Advance online publication. https://doi.org/10.1007/s43440-020-00203-1

Wei Y, Chang L, Hashimoto K (2020) A historical review of antidepressant effects of ketamine and its enantiomers. Pharmacol Biochem Behav 190:172870

Wei Y, Chang L, Hashimoto K (2021) Molecular mechanisms underlying the antidepressant actions of arketamine: beyond the NMDA receptor. Mol Psychiatry 2021 May 7. https://doi.org/10.1038/s41380-021-01121-1.

Xie ZM, Wang XM, Xu N, Wang J, Pan W, Tang XH, Zhou ZQ, Hashimoto K, Yang JJ (2017) Alterations in the inflammatory cytokines and brain-derived neurotrophic factor contribute to depression-like phenotype after spared nerve injury: improvement by ketamine. Sci Rep 7(1):3124

Xiong Z, Fujita Y, Zhang K, Pu Y, Chang L, Ma M, Chen J, Hashimoto K (2019) Beneficial effects of (*R*)-ketamine, but not its metabolite (2*R*,6*R*)-hydroxynorketamine, in the depression-like phenotype, inflammatory bone markers, and bone mineral density in a chronic social defeat stress model. Behav Brain Res 368:111904

Yang C, Kobayashi S, Nakao K, Dong C, Han M, Qu Y, Ren Q, Zhang JC, Ma M, Toki H, Yamaguchi JI, Chaki S, Shirayama Y, Nakazawa K, Manabe T, Hashimoto K (2018a) AMPA receptor activation-independent antidepressant actions of ketamine metabolite (*S*)-Norketamine. Biol Psychiatry 84(8):591–600

Yang C, Ren Q, Qu Y, Zhang JC, Ma M, Dong C, Hashimoto K (2018b) Mechanistic target of rapamycin-independent antidepressant effects of (*R*)-ketamine in a social defeat stress model. Biol Psychiatry 83(1):18–28

Yang C, Shirayama Y, Zhang JC, Ren Q, Yao W, Ma M, Dong C, Hashimoto K (2015a) *R*-ketamine: a rapid-onset and sustained antidepressant without psychotomimetic side effects. Transl Psychiatry 5(9):e632

Yang C, Wardenaar KJ, Bosker FJ, Li J, Schoevers RA (2019a) Inflammatory markers and treatment outcome in treatment resistant depression: a systematic review. J Affect Disord 257:640–649

Yang C, Yang J, Luo A, Hashimoto K (2019b) Molecular and cellular mechanisms underlying the antidepressant effects of ketamine enantiomers and its metabolites. Transl Psychiatry 9(1):280

Yang JJ, Wang N, Yang C, Shi JY, Yu HY, Hashimoto K (2015b) Serum interleukin-6 is a predictive biomarker for ketamine's antidepressant effect in treatment-resistant patients with major depression. Biol Psychiatry 77(3):e19–e20

Zhan Y, Zhou Y, Zheng W, Liu W, Wang C, Lan X, Deng X, Xu Y, Zhang B, Ning Y (2020) Alterations of multiple peripheral inflammatory cytokine levels after repeated ketamine infusions in major depressive disorder. Transl Psychiatry 10(1):246

Zhang JC, Li SX, Hashimoto K (2014) *R*(−)-ketamine shows greater potency and longer lasting antidepressant effects than *S*(+)-ketamine. Pharmacol Biochem Behav 116:137–141

Zhang JC, Yao W, Hashimoto K (2016) Brain-derived neurotrophic factor (BDNF)-TrkB signaling in inflammation-related depression and potential therapeutic targets. Curr Neuropharmacol 14(7):721–731

Zhang K, Hashimoto K (2019) An update on ketamine and its two enantiomers as rapid-acting antidepressants. Expert Rev Neurother 19(1):83–92

Zhang K, Ma M, Dong C, Hashimoto K (2018) Role of inflammatory bone markers in the antidepressant actions of (*R*)-ketamine in a chronic social defeat stress model. Int J Neuropsychopharmacol 21(11):1025–1030

Zhang K, Yang C, Chang L, Sakamoto A, Suzuki T, Fujita Y, Qu Y, Wang S, Pu Y, Tan Y, Wang X, Ishima T, Shirayama Y, Hatano M, Tanaka KF, Hashimoto K (2020) Essential role of microglial transforming growth factor-β1 in antidepressant actions of (*R*)-ketamine and the novel antidepressant TGF-β1. Transl Psychiatry 10(1):32

Zheng W, Zhou YL, Wang CY, Lan XF, Zhang B, Yang MZ, Nie S, Ning YP (2020) Neurocognitive effects of six ketamine infusions and the association with antidepressant effects in treatment-resistant bipolar depression: a preliminary study. PeerJ 8:e10208

Zhong M, Wang X, Xiao J, Yi J, Zhu X, Liao J, Wang W, Yao S (2011) Amygdala hyperactivation and prefrontal hypoactivation in subjects with cognitive vulnerability to depression. Biol Psychol 88(2–3):233–242

Zhou Y, Zheng W, Liu W, Wang C, Zhan Y, Li H, Chen L, Li M, Ning Y (2018a) Antidepressant effect of repeated ketamine administration on kynurenine pathway metabolites in patients with unipolar and bipolar depression. Brain Behav Immun 74:205–212

Zhou Y, Zheng W, Liu W, Wang C, Zhan Y, Li H, Chen L, Li M, Ning Y (2018b) Neurocognitive effects of six ketamine infusions and the association with antidepressant response in patients with unipolar and bipolar depression. J Psychopharmacol 32(10):1118–1126

Zhou YL, Wu FC, Liu WJ, Zheng W, Wang CY, Zhan YN, Lan XF, Ning YP (2020a) Volumetric changes in subcortical structures following repeated ketamine treatment in patients with major depressive disorder: a longitudinal analysis. Transl Psychiatry 10(1):264

Zhou YL, Wu FC, Wang CY, Zheng W, Lan XF, Deng XR, Ning YP (2020b) Relationship between hippocampal volume and inflammatory markers following six infusions of ketamine in major depressive disorder. J Affect Disord 276:608–615

Nitrous Oxide: An Old Compound with Emerging Psychotropic Properties

Lojine Y. Kamel, Darin F. Quach, Britt M. Gott, and Charles R. Conway

Abstract Nitrous oxide has been utilized in medicine for over 150 years, with current applications in emergency medicine, dentistry, and obstetrics. It is a commonly utilized anesthetic and analgesic that is well-tolerated among adult users. Outside of medicine, nitrous oxide is a frequently inhaled recreational drug, typically used among teenagers and young adults. Within the last decade, nitrous oxide has been found to have antidepressant effects similar to ketamine, thought to be related to its effect on N-methyl-D-aspartate receptors (NMDARs). Though there is limited evidence about its therapeutic effects in mood disorders, recent studies support that nitrous oxide has the potential to treat resistant unipolar depression. In this chapter, we discuss nitrous oxide's putative mechanism of action, administration, current use in psychiatry, and possible future psychotropic applications.

Keywords Nitrous Oxide · Treatment Resistant Major Depression (TRMD) · Antidepressant · N-methyl-D-aspartate receptor (NMDAR) · Ketamine · Suicide Treatments

Brief Medical History of Nitrous Oxide

Nitrous oxide is a colorless nonflammable gas that has been utilized in the medical field for the last 150 years. It was first artificially synthesized by the English chemist Joseph Priestley in 1772, who referred to it as "inflammable nitrous air." In 1799, Cornish chemist Humphry Davy nicknamed it "laughing gas" after observing how inhaling the gas induced laughter. A year later, he became the first person to write about nitrous oxide's anesthetic and analgesic potential in his book: *Researches, Chemical and Philosophical* (1800).

In 1844, American dentist Horace Wells was the first to utilize nitrous oxide for its analgesic properties in a medical setting, first testing the gas on himself for a

L. Y. Kamel · D. F. Quach · B. M. Gott · C. R. Conway (✉)
Washington University School of Medicine, St Louis, MO, USA
e-mail: conwaycr@wustl.edu

K. Hashimoto, M. Manto (eds.), *New Rapid-acting Antidepressants*, Contemporary Clinical Neuroscience,
https://doi.org/10.1007/978-3-030-79790-4_4

tooth extraction and then successfully employing it in 12 dental procedures (Lew et al. 2018). Nitrous oxide was not popularly adopted among dentists until two decades later, when former medical student Gardner Quincy Colton opened the Colton Dental Association. The clinic ultimately utilized nitrous oxide as their primary analgesic agent and successfully treated over 25,000 patients between 1864 and 1897 (Miller 1941). Clinicians soon realized that inhaling pure nitrous oxide could cause hypoxia, which could be prevented by administering the gas with oxygen. In 1911, American anesthetist Arthur Ernest Guedel became the first to write about the combination of nitrous oxide and oxygen utilized as a self-administered anesthetic (Podwińska et al. 2008). Eventually, a mixture of nitrous oxide and oxygen became a staple agent utilized in dental anesthesia, and it is still frequently employed in this capacity today.

Current Medical Applications of Nitrous Oxide

Today, nitrous oxide is still commonly utilized as a mild inhaled anesthetic and anxiolytic in a variety of short procedures in emergency medicine, obstetrics, and dentistry. Used in procedures requiring mild, moderate, and deep sedation, nitrous oxide is often paired with an additional anesthetic agent to achieve deeper degrees of sedation (Becker and Rosenberg 2008). In addition to its analgesic use, nitrous oxide has the added advantage of having anxiolytic properties and its associated ability to suppress gag and swallow reflexes (Wilson 2013a; b), which makes it particularly well suited for dental procedures.

On top of its extensive use in dentistry (an estimated 70% of dental practices use nitrous oxide-oxygen mixtures to achieve patient sedation), the use of inhaled nitrous has found favor in other medical settings (American Dental Association 2008). In the emergency room, inhaled nitrous oxide is frequently used in laceration repair (Burton et al. 1998), fracture reduction (Hennrikus et al. 1994), venipuncture (Gerhardt et al. 2001), and other brief procedures. In obstetrics, a blend of nitrous oxide and oxygen is used as a form of analgesia during labor and post-delivery procedures like manual placenta removal, uterine curettage, and laceration repair (Bishop 2007). Most recently, nitrous oxide has been found to have considerable potential as an antidepressant (see below).

Nitrous Oxide as an Antidepressant

Nitrous oxide is an N-methyl-D-aspartate receptor (NMDAR) antagonist. Emerging from the recent success of another rapid-acting NMDAR antagonist, ketamine, Nagele et al. studied inhaled nitrous oxide with oxygen in a double-blind, placebo-controlled, crossover trial (Nagele et al. 2015). In this trial, 20 patients suffering from treatment-resistant major depression (TRMD) were given 1 hour treatments of either active agent (50–50% nitrous-oxide-oxygen mixture) or placebo and assessed for antidepressant response over the course of 24 hours. For this trial, TRMD was carefully defined as failing at least two adequate dose-duration antidepressant trials. Additionally, care was taken to exclude potential confounding diagnoses to TRMD, including substance use disorders, severe personality disorder, or any history of bipolar or psychotic illness. In this trial, the administration of 50% nitrous oxide over 60 minutes resulted in a significant improvement in depression scale scores measured by the Hamilton Depression Rating Scale. Compared to placebo, this was statistically significant at both 2 and 24 hours post-treatment with an overall response rate of 20% (4 out of 20) and three individuals with a complete remission of symptoms; one placebo patient responded and none remitted. These findings were promising, as this group of TRMD patients were severely resistant (mean number of failed antidepressant trials ~8). The effects of the treatment, similar to what has been observed with ketamine, were more sustained than anticipated, which complicated analyses of the crossover, as the two treatment exposures were separated only by 1 week.

Overall, the treatment was relatively well-tolerated with no evidence of psychosis. The most prevalent side effect was nausea and vomiting, which was observed in three individuals. This was followed by numbness/paresthesia and anxiety, which was experienced by two individuals in each category. There was no increase in plasma total homocysteine following nitrous oxide treatment suggesting minimal inactivation of vitamin B12-dependent metabolism. Of note, no structured scales to assess adverse effects were used, making it possible that subtle side effects were missed.

Interestingly, there were four depressive symptoms including depressed mood, guilt, suicidal ideation, and anxiety that appeared to be more responsive to nitrous oxide, as they demonstrated the largest positive change from baseline. Although this requires additional investigation, these results suggest that nitrous oxide may be targeting specific depressive symptomatology.

Putative Mechanisms of Nitrous Oxide as an Antidepressant

While there is a long history of nitrous oxide use as an anesthetic and analgesic, its utility in psychiatry is relatively nascent. Moreover, the mechanism of action as an antidepressant remains largely unknown. Drawing from preclinical studies in tissue culture and rodent models, nitrous oxide has been shown to act on various receptors.

One of the most studied mechanisms involves NMDA receptors, where nitrous oxide acts as an antagonist (refer to Fig. 1). Nitrous oxide has been demonstrated to have protective effects against excitotoxicity elicited by NMDA agonists, N-methyl-D,L aspartic acid, and MK0801, in adult female rats (Jevtović-Todorović et al. 1998). In that study, direct electrophysiologic measurements were performed in cultured hippocampus neurons where nitrous oxide inhibited currents elicited by NMDA. At saturating concentrations, NMDA was not able to overcome inhibition

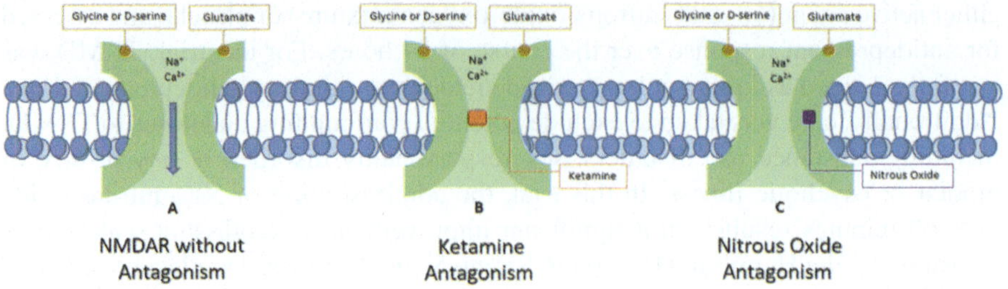

Fig. 1 N-methyl-D-aspartate receptors (NMDARs) are a type of ion channel transmembrane receptor, composed of 4 subunits (only 2 depicted here), which act to allow selective entry of cations (particularly Na^+ and Ca^{2+}). In order for cations to pass, NMDARs must have simultaneous binding of glutamate and a co-agonist (glycine or D-serine). Additionally, endogenous ions and modulators can bind to NMDARs to alter ion channel function, including Mg^{2+}, Zn^{2+}, H^+, polyamines, neurosteroids, and oxysterols. NMDARs exhibit excitatory synaptic function with a combination of presynaptic (glutamate release) and postsynaptic activation (depolarization). Panel (**A**) depicts the NMDAR in its active state; typically in this state there is cation passage. Panel (**B**) depicts NMDAR antagonism with exposure to ketamine. Ketamine is a noncompetitive NMDAR antagonist; the ketamine binding site is believed to be within the membrane channel. When NMDAR ion channels are open, ketamine enters the channel, binds to a site within its pore, and prevents ions from passing through it. The ion channel is able to close around ketamine creating a "trapping ion channel block." The channel can be reopened when glutamate (or another agonist) binds to the receptor and the membrane is depolarized. Panel (**C**) depicts the binding of nitrous oxide to the NMDAR. Nitrous oxide is also a noncompetitive NMDAR antagonist. Unlike ketamine, nitrous oxide's effects are weakly voltage-dependent. Little is known about its true mechanism, and the binding site location has yet to be fully elucidated; however, existing studies support that it is not a trapping ion channel blocker and likely acts allosterically by binding to a portion of the receptor outside the ion channel (Zorumski et al. 2015)

by nitrous oxide, indicating that nitrous was acting through a noncompetitive mechanism. Subsequent studies have established the requirement of NMDA receptors for certain actions of nitrous oxide (Nagele et al. 2004; Sato et al. 2005).

Related to NMDA, alpha-amino-3-hydroxy-5-methyl-4-isoxazole propionic acid (AMPA)/kainite-type glutamate receptors are also weakly inhibited by nitrous oxide (Mennerick et al. 1998). Actions at AMPA receptors are of interest because activation of these receptors by ketamine has been shown to increase BDNF production and stimulate neurogenesis in the hippocampus in preclinical models (Jourdi et al. 2009; Kyoung et al. 2020). The limited clinical studies investigating AMPA receptor activity have involved the use of ketamine, and it is unknown whether nitrous has similar effects. In a preclinical murine model, Zanos et al., using an AMPA receptor antagonist, demonstrated that the antidepressant effects of ketamine require AMPA receptor activation (Zanos et al. 2016). Further, in two separate studies, one with healthy subjects and another comparing healthy subjects with individuals diagnosed with major depressive disorder, administration of ketamine was demonstrated to increase glutaminergic signaling in regions with increased levels of AMPA receptors (Li et al. 2017). Additionally, improvements in depression ratings correlated with reduced NMDA and AMPA connectivity within a somatosensory cortical network (Gilbert et al. 2018). As these studies were performed employing ketamine, future studies using nitrous oxide will be needed to determine if there are similar AMPA-mediated effects contributing to nitrous oxide's antidepressant effects.

Interestingly, some anesthetic and analgesic effects mediated by nitrous oxide are attributable to its activity on opioid receptors (Koyama and Fukuda 2010; Fukagawa et al. 2014). Given that kappa opioid receptors have been shown to potentially play a role in antidepressant effects, and blockade of opioid receptors can attenuate the antidepressant effects of ketamine, it is possible that nitrous oxide may also act through opioid receptors in a similar manner (Falcon et al. 2016; Williams et al. 2018).

Nitrous oxide also has effects on other ion channels including low voltage activated calcium channels, gamma-aminobutyric acid receptors A and C, 5-hydroxytryptamine 3, and glycine receptors. Nitrous oxide additionally has effects on membrane fluidity, akin to some other anesthetics, but it is unclear whether these effects contribute to effects on ion channels. Moreover, it is unclear whether effects of nitrous oxide on ion channels other than NMDA receptors contribute to antidepressant effects (Downie et al. 1996; Daniels and Roberts 1998; Mennerick et al. 1998; Yamakura and Harris 2000; Todorovic et al. 2001).

Studies investigating downstream signaling pathways have demonstrated that exposure to nitrous oxide leads to hippocampal cell proliferation and activation of neuronal nitric oxide synthase (nNOS) in rodent models (Chamaa et al. 2018; Liu et al. 2020). Furthermore, activation of nNOS in the medial prefrontal cortex led to increased expression levels of brain-derived neurotrophic factor and antidepressant-like effects in mice (Liu et al. 2020).

Thus far, human studies attempting to understand the mechanisms of action of nitrous have been limited. Early studies using electroencephalography (EEG) suggest that administration of nitrous oxide decreases functional connectivity in the superficial parietal network (Kuhlmann, Foster and Liley 2013). Additionally, frontal slow wave (delta) activity is modulated by nitrous oxide (Foster and Liley 2013; Pavone et al. 2016). However, the significance of these findings with respect to possible antidepressant effects of nitrous oxide remains largely unknown.

Administration of Nitrous Oxide

Though nitrous oxide is noncombustible and nonflammable, it has the capability to support combustion and should thus not be stored near open flames. It is customarily stored under pressure in color-coded tanks and is typically self-administered through a demand-valve mask or mouthpiece as a 50–50% or 70–30% nitrous oxide-oxygen mixture. Acute intoxication symptoms include rapid onset of excitement, dissociation, and euphoria, which resolve within minutes. Though nitrous oxide is odorless, the tubing and mask used during administration can often have an unpleasant smell, particularly for children. The insides of pediatric masks are thus frequently coated with a pleasant smelling liquid such as strawberry or bubblegum (Huang and Johnson 2016).

The most commonly noted side effects of nitrous oxide include nausea, vomiting, headache, tingling, dizziness, and euphoria. Longer duration of sedation typically increases the risk of side effects, particularly nausea and vomiting (Peyton and Wu 2014). Chronic and recurrent exposure to nitrous oxide, as occurs in nitrous oxide abusers, has been linked to adverse effects such as infertility, spontaneous abortion, neurologic deficits, vitamin B12 deficiency, and blood dyscrasias.

While nitrous oxide use is relatively safe, it is not without potential serious side effects, which can include oxygen desaturation, bradycardia, or aspiration (Becker and Rosenberg 2008). These major complications are very rare and found to be more common in younger pediatric patients and when combined with specific agents (e.g., benzodiazepines) (Gall et al. 2001).

Though safe to use for the vast majority of patients, there are several absolute and relative contraindications to nitrous oxide use. Absolute contraindications include inability to use a nasal face mask, either due to physical (i.e., nasopharyngeal obstructions, upper respiratory infection, severe sinusitis) or psychological/cognitive reasons (i.e., phobia, cognitive impairment, uncooperative), unconsciousness, gastrointestinal distention, pneumothorax, and intracranial hypertension (Becker and Rosenberg 2008). Relative contraindications vary, but may include administration of nitrous oxide to children under 4 years or use on children with underlying conditions that might affect respiratory or brain function (Gall et al. 2001).

Dosing of Nitrous Oxide in Depression

For use in treatment-resistant major depression, previous studies (Nagele et al. 2015, 2020) have targeted a 50–50% nitrous-oxide-oxygen mixture dosage for one hour treatments, as mentioned previously. Although the total number of participants across both studies is small ($N = 21$), the reported side effect profile was relatively mild and short-lived (Nagele et al. 2015). As demonstrated in previous animal studies {Jevtović-Todorović et al. 1998 #2208}, NMDAR antagonism can occur at nitrous oxide concentrations below 50%, and this observation has prompted dose-finding studies in patient groups with TRMD. To elucidate effective dosage and determine dose-dependent side effect occurrence and duration of antidepressant response, a trial has been completed by the same group as the Nagele et al. (2015) study. Using a randomized, double-blind, crossover design, this trial compared three doses of nitrous oxide: 50% nitrous oxide-oxygen mixture, 25% nitrous oxide-oxygen mixture, and a placebo with 0% nitrous oxide-oxygen mixture. TRMD participants received hour-long inhalation sessions to each of the three concentrations, separated by at least 4 weeks. As of the beginning of 2021, the results of this trial are not yet published.

Administration of Nitrous Oxide in Depression

Administration of nitrous oxide should occur via a standard anesthesia face mask. Critical considerations of nitrous oxide delivery include ensuring the system does not have or develop leaks in gas delivery and also ensuring the adequate ventilation through a wall-suction gas scavenging system. Suction and airway equipment should be readily available. Delivery should be monitored by appropriately trained staff in administering sedatives and airway support, per the American Society of Anesthesia guidelines on sedation and analgesia, through nitrous oxide and oxygen/air flow meters. Patient safety should be maintained through infrared pulse oximeters for detecting potential hypoxemia. Additional monitoring by continuous three-lead ECG, noninvasive blood pressure, and end-tidal CO_2 under the supervision of an attending-level anesthesiologist may be warranted, as well as physician-approved discharge following a short observatory period.

Nitrous Oxide Abuse Potential

After Humphry Davy wrote about "laughing gas" in his abovementioned book, nitrous oxide quickly became a popular recreational drug among the British upper class, who frequently held "laughing gas parties" in which they would pass around a silk bag filled with the gas. Though nitrous oxide became increasingly more popular as a recreational drug in the nineteenth and twentieth centuries, many countries restricted its purchase to strictly medical purposes due to its limited availability (Randhawa and Bodenham 2016). Despite this, use of nitrous oxide for pleasure continued, and it is still commonly used as a recreational inhalant to this day.

Currently, recreational nitrous oxide is often referred to by its more popular names including "laughing gas," "whippets," "hippy crack," and "nagging" (Garakani et al. 2016). Administration typically involves inhaling the gas from bulbs or balloons, or cartridges commonly used for whipped cream dispensation (hence the term "whippets"). The gas is primarily used as a drug of abuse by teenagers and young adults, as it can be obtained cheaply and easily. In the United States, 9 percent of 12- to 17-year-olds have used inhalants, 21% of whom have reported using nitrous oxide at least once. Nitrous oxide is thus the fourth most commonly used inhalant in the United States among this age group (Wu et al., 2004).

Despite this, nitrous oxide's addiction potential is minimal, as frequent users typically do not develop physical or psychological dependence on it (van Amsterdam, Nabben & van den Brink, 2015). As mentioned above, regular users may experience nitrous oxide toxicity with chronic use, and there have been few case reports of death related to nitrous oxide inhalation (Garriott and Petty 1980; Bowen et al. 1999).

Nitrous Oxide Compared to Ketamine

Further supporting the importance of the NMDAR in depression is the emergence of ketamine as a rapid-acting, antidepressant treatment. Though both nitrous oxide and ketamine are known NMDAR antagonists, they have distinctly different properties summarized in Table 1.

Ketamine is a noncompetitive NMDAR antagonist that has been used as an analgesic and anesthetic since the 1970s. More recently, ketamine has been found to have rapid-acting antidepressant properties, and the inhaled version, esketamine, has received FDA-approved indications for both treatment-resistant depression and suicidal ideation occurring in the context of depression (Canuso et al. 2018; Swainson et al. 2019).

Research examining ketamine's putative antidepressant effects began in the 1990s, with initial findings suggesting its potential to improve depression in animal models (Papp and Moryl 1994; Trullas and Skolnick 1990). In 2000, the first clinical trial utilizing IV ketamine for depression in humans resulted in significant improvement in depression 72 hours after administration of just one dose (0.5 mg/kg) (Berman et al. 2000). Ongoing research has subsequently supported this initial finding, with more recent studies describing IV ketamine's acute antidepressant effect in patients with unipolar depression (Murrough et al. 2013a; b; Zarate Jr et al. 2006), bipolar depression (DiazGranados et al. 2010a; b; Zarate et al. 2012), and suicidality (DiazGranados et al. 2010a, b; Grunebaum et al., 2018) and TRMD patients who had previously failed ECT (Ibrahim et al. 2011). Further studies have demonstrated rapid and prolonged antidepressant effects of ketamine with multiple doses (Murrough et al. 2013a; b; Singh et al. 2016). Finally, the more active S-enantiomer of ketamine, esketamine, has more recently been the target of antidepressant studies due to its propensity to benefit mood with comparatively minimal side effects (Pfenninger et al. 2002; Canuso et al. 2018).

Table 1 Comparison of NMDAR antagonists ketamine and nitrous oxide

	Ketamine	Nitrous oxide
NMDAR aspects		
NMDAR noncompetitive antagonist	x	x
Voltage dependent	x	Weakly
Open channel blocker	x	
Effects on AMPARs	None	Weak antagonist
Commonly observed side effects		
Elevated blood pressure	x	
Nausea		x
Psychosis	x	
Euphoria (during exposure)	x	x
Other		
Controlled substance	x	
Abuse potential	High	Limited

Though several therapeutic trials of nitrous oxide in TRMD are currently underway, at present, there is limited evidence of nitrous oxide's potential to treat mood disorders. As described previously, Nagele et al. (2015) published the first double-blind trial demonstrating antidepressant efficacy in TRMD. Nagele et al. (2020) also described a case of patient with a 15-year history of severe recurrent TRMD who achieved and maintained remission for over a month with a single dose of nitrous oxide. A study by Das et al. additionally explored nitrous oxide's promising ability to mitigate psychological trauma (2016). Further research examining nitrous oxide's durability and capability to treat other mood and other psychiatric disorders is currently underway (Nagele et al. 2020).

Though the evidence of antidepressant efficacy currently is not as strong as ketamine, nitrous oxide has several potential therapeutic advantages over ketamine. Ketamine has significant abuse and addictive potential; hence, the US Drug Enforcement Agency lists ketamine as a schedule III drug. In contrast, nitrous oxide, though having some abuse potential, is not believed to be highly addictive and does not have DEA classification as a scheduled drug. Further, to date, there is no evidence that nitrous oxide induces psychosis, which is a known potential side effect of ketamine (Lahti et al. 2001; Bowdle et al. 1998). Further, nitrous oxide is not associated with elevated blood pressure, as has been observed with ketamine (Wan et al. 2015). Finally, though nitrous oxide and ketamine both have an acute onset of action, nitrous oxide's rapid elimination allows for less restrictions for patients post-administration – dental patients routinely drive themselves home after receiving nitrous oxide; inhaled ketamine (for TRMD) currently prohibits patients from driving themselves home following treatment (Giorgetti et al. 2015a; b). As nitrous oxide research increases, a true comparison of ketamine and nitrous oxide's antidepressant properties and side effect profiles will become more clear.

Future Directions

Though current nitrous oxide research has demonstrated evidence to support that this compound has antidepressant and possibly anti-suicidal properties, many questions about its efficacy and therapeutic potential remain. Several outstanding questions include: What is the effective dose of nitrous oxide required for antidepressant effects? Previous animal studies suggest that dosing below the 50% used in dental anesthesia may bring about adequate NMDAR antagonism (Jevtović-Todorović et al. 1998). Further, what is considered an adequate duration of nitrous oxide exposure? As noted above, very little is understood about how nitrous oxide affects the brain, both in non-depressed patients and TRMD patients. Hence, imaging and electroencephalographic (EEG) studies may provide some clarification of these effects. As there is currently a limited breadth of research examining the psychiatric therapeutic potential of nitrous oxide, further studies are needed to ascertain if nitrous oxide has the additional potential to treat other types of psychiatric disorders such as post-traumatic stress disorder, bipolar disorder, anxiety, or addiction (Das et al.,

2016); Dimick et al., 2020). Given its relative tolerability, ease of administration, history of clinical use, and potential ability to treat several psychiatric conditions, it is feasible that nitrous oxide may soon become a popularly prescribed agent by psychiatrists. Like ketamine, there may be utility in the creation of specialty nitrous oxide clinics, whereby patients may receive doses of the agent under medical supervision.

References

American Dental Association Survey Center 2007 (2008) Survey of current issues in dentistry: surgical dental implants, amalgam restorations, and sedation. American Dental Association

Becker DE, Rosenberg M (2008) Nitrous oxide and the inhalation anesthetics. Anesth Prog 55(4). https://doi.org/10.2344/0003-3006-55.4.124

Berman RM et al (2000) Antidepressant effects of ketamine in depressed patients. Biol Psychiatry 47(4):351–354. https://doi.org/10.1016/S0006-3223(99)00230-9

Bishop JT (2007) Administration of Nitrous Oxide in labor: expanding the options for women. J Midwifery Womens Health 52(3):308–309. https://doi.org/10.1016/j.jmwh.2007.02.018

Bowdle TA et al (1998) Psychedelic effects of ketamine in healthy volunteers: relationship to steady-state plasma concentrations. Anesthesiology 88(1):82–88. https://doi.org/10.1097/00000542-199801000-00015

Bowen SE, Daniel J, Balster RL (1999) Deaths associated with inhalant abuse in Virginia from 1987 to 1996. Drug Alcohol Depend 53(3):239–245. https://doi.org/10.1016/S0376-8716(98)00139-2

Burton JH, Auble TE, Fuchs SM (1998) Effectiveness of 50% nitrous oxide/50% oxygen during laceration repair in children. Acad Emerg Med 5(2):112–117. https://doi.org/10.1111/j.1553-2712.1998.tb02594.x

Canuso CM et al (2018) Efficacy and safety of intranasal esketamine for the rapid reduction of symptoms of depression and suicidality in patients at imminent risk for suicide: results of a double-blind, randomized, placebo-controlled study. Am J Psychiatr 175(7):620–630. https://doi.org/10.1176/appi.ajp.2018.17060720

Chamaa F et al (2018) Nitrous oxide induces prominent cell proliferation in adult rat hippocampal dentate gyrus. Front Cell Neurosci 12(May):1–8. https://doi.org/10.3389/fncel.2018.00135

Daniels S, Roberts RJ (1998) Post-synaptic inhibitory mechanisms of anaesthesia; glycine receptors. Toxicol Lett 100–101:71–76. https://doi.org/10.1016/S0378-4274(98)00167-2

Das, R. K., Tamman, A., Nikolova, V., Freeman, T. P., Bisby, J. A., Lazzarino, A. I., & Kamboj, S. K. (2016). Nitrous oxide speeds the reductionof distressing intrusive memories in an experimental model of psychological trauma. Psychological medicine, 46(8), 1749–1759. https://doi.org/10.1017/S003329171600026X

DiazGranados N et al (2010a) Rapid resolution of suicidal ideation after a single infusion of an N-methyl-D-aspartate antagonist in patients with treatment-resistant major depressive disorder. J Clin Psychiatry 71(12):1605–1611. https://doi.org/10.4088/JCP.09m05327blu

Diazgranados N et al (2010b) A randomized add-on trial of an N-methyl-D-aspartate antagonist in treatment-resistant bipolar depression. Arch Gen Psychiatry 67(8):793–802. https://doi.org/10.1001/archgenpsychiatry.2010.90

Dimick, M. K., Omrin, D., MacIntosh, B. J., Mitchell, R., Riegert, D., Levitt, A., Schaffer, A., Belo, S., Iazzetta, J., Detzler, G., Choi, M., Choi, S., Orser, B. A., & Goldstein, B. I. (2020). Nitrous oxide as a putative novel dual-mechanism treatment for bipolar depression: Proof-of-concept study design and methodology. Contemporary clinical trials communications, 19, 100600. https://doi.org/10.1016/j.conctc.2020.100600

Downie DL et al (1996) Effects of inhalational general anaesthetics on native glycine receptors in rat medullary neurones and recombinant glycine receptors in Xenopus oocytes. Br J Pharmacol 118(3):493–502. https://doi.org/10.1111/j.1476-5381.1996.tb15430.x

Falcon E et al (2016) Antidepressant-like effects of buprenorphine are mediated by kappa opioid receptors. Neuropsychopharmacology 41(9):2344–2351. https://doi.org/10.1038/npp.2016.38

Foster BL, Liley DTJ (2013) Effects of nitrous oxide sedation on resting electroencephalogram topography. Clin Neurophysiol 124(2):417–423. https://doi.org/10.1016/j.clinph.2012.08.007

Fukagawa H, Koyama T, Fukuda K (2014) κ-Opioid receptor mediates the antinociceptive effect of nitrous oxide in mice. Br J Anaesth 113(6):1032–1038. https://doi.org/10.1093/bja/aeu254

Gall O et al (2001) Adverse events of premixed nitrous oxide and oxygen for procedural sedation in children. Lancet 358(9292):1514–1515. https://doi.org/10.1016/S0140-6736(01)06575-8

Garakani A et al (2016) Neurologic, psychiatric, and other medical manifestations of nitrous oxide abuse: a systematic review of the case literature. Am J Addict 25(5):358–369. https://doi.org/10.1111/ajad.12372

Garriott J, Petty CS (1980) Death from inhalant abuse: toxicological and pathological evaluation of 34 cases. Clin Toxicol 16(3):305–315. https://doi.org/10.3109/15563658008989954

Gerhardt RT, King KM, Wiegert RS (2001) Inhaled nitrous oxide versus placebo as an analgesic and anxiolytic adjunct to peripheral intravenous cannulation. Am J Emerg Med 19(6):492–494. https://doi.org/10.1053/ajem.2001.25780

Gilbert JR et al (2018) Glutamatergic signaling drives ketamine-mediated response in depression: evidence from dynamic causal modeling. Int J Neuropsychopharmacol 21(8):740–747. https://doi.org/10.1093/ijnp/pyy041

Giorgetti R et al (2015a) Effects of ketamine on psychomotor, sensory and cognitive functions relevant for driving ability. Forensic Sci Int 252:127–142. https://doi.org/10.1016/j.forsciint.2015.04.024

Giorgetti R et al (2015b) Effects of ketamine on psychomotor, sensory and cognitive functions relevant for driving ability. Forensic Sci Int 252:127–142. https://doi.org/10.1016/j.forsciint.2015.04.024

Grunebaum, M. F., Galfalvy, H. C., Choo, T. H., Keilp, J. G., Moitra, V. K., Parris, M. S., Marver, J. E., Burke, A. K., Milak, M. S., Sublette, M. E., Oquendo, M. A., & Mann, J. J. (2018). Ketamine for Rapid Reduction of Suicidal Thoughts in Major Depression: A Midazolam-Controlled Randomized Clinical Trial. The American journal of psychiatry, 175(4), 327–335. https://doi.org/10.1176/appi.ajp.2017.17060647

Hennrikus WL et al (1994) Self-administered nitrous oxide analgesia for pediatric fracture reductions. J Pediatr Orthop 14(4):538–542. https://doi.org/10.1097/01241398-199407000-00023

Huang C, Johnson N (2016) Nitrous oxide, from the operating room to the emergency department. Current Emergency and Hospital Medicine Reports 4(1):11–18. https://doi.org/10.1007/s40138-016-0092-3

Ibrahim L et al (2011) Rapid decrease in depressive symptoms with an N-methyl-d-aspartate antagonist in ECT-resistant major depression. Prog Neuro-Psychopharmacol Biol Psychiatry 35(4):1155–1159. https://doi.org/10.1016/j.pnpbp.2011.03.019

Jevtović-Todorović V et al (1998) Nitrous oxide (laughing gas) is an NMDA antagonist, neuroprotectant and neurotoxin. Nat Med 4(4):460–463. https://doi.org/10.1038/nm0498-460

Jourdi H et al (2009) Positive AMPA receptor modulation rapidly stimulates BDNF release and increases dendritic mRNA translation. J Neurosci 29(27):8688–8697. https://doi.org/10.1523/JNEUROSCI.6078-08.2009

Koyama T, Fukuda K (2010) Involvement of the κ-opioid receptor in nitrous oxide-induced analgesia in mice. J Anesth 24(2):297–299. https://doi.org/10.1007/s00540-010-0886-5

Kuhlmann L, Foster BL, Liley DTJ (2013) Modulation of functional EEG networks by the NMDA antagonist nitrous oxide. PLoS One 8(2). https://doi.org/10.1371/journal.pone.0056434

Lahti AC et al (2001) Effects of ketamine in normal and schizophrenic volunteers. Neuropsychopharmacology 25(4):455–467. https://doi.org/10.1016/S0893-133X(01)00243-3

Lew V, McKay E, Maze M (2018) Past, present, and future of nitrous oxide. Br Med Bull 125(1):103–119. https://doi.org/10.1093/bmb/ldx050

Li M et al (2017) Temporal dynamics of antidepressant ketamine effects on glutamine cycling follow regional fingerprints of AMPA and NMDA receptor densities. Neuropsychopharmacology 42(6):1201–1209. https://doi.org/10.1038/npp.2016.184

Liu W et al (2020) Repeated nitrous oxide exposure exerts antidepressant-like effects through neuronal nitric oxide synthase activation in the medial prefrontal cortex. Front Psych 11(September):1–11. https://doi.org/10.3389/fpsyt.2020.00837

Mennerick S et al (1998) Effect of nitrous oxide on excitatory and inhibitory synaptic transmission in hippocampal cultures. J Neurosci 18(23):9716–9726. https://doi.org/10.1523/jneurosci.18-23-09716.1998

Miller AH (1941) Technical development of gas anesthesia. Anesthesiology 2(4):398–409. https://doi.org/10.1097/00000542-194107000-00004

Murrough JW et al (2013a) Antidepressant efficacy of ketamine in treatment-resistant major depression: a two-site randomized controlled trial. Am J Psychiatr 170(10):1134–1142. https://doi.org/10.1176/appi.ajp.2013.13030392

Murrough JW et al (2013b) Rapid and longer-term antidepressant effects of repeated ketamine infusions in treatment-resistant major depression. Biol Psychiatry 74(4):250–256. https://doi.org/10.1016/j.biopsych.2012.06.022

Nagele P et al (2020) Prolonged remission of major depressive disorder after single nitrous oxide inhalation treatment. Front Psych 11(July):10–11. https://doi.org/10.3389/fpsyt.2020.00692

Nagele P et al (2015) Nitrous oxide for treatment-resistant major depression: a proof-of-concept trial. Biol Psychiatry 78(1):10–18. https://doi.org/10.1016/j.biopsych.2014.11.016

Nagele P, Metz LB, Crowder CM (2004) Nitrous oxide (N2O) requires the N-methyl-D-aspartate receptor for its action in Caenorhabditis elegans. Proc Natl Acad Sci U S A 101(23):8791–8796. https://doi.org/10.1073/pnas.0402825101

Papp M, Moryl E (1994) Antidepressant activity of non-competitive and competitive NMDA receptor antagonists in a chronic mild stress model of depression. Eur J Pharmacol 263(1–2):1–7. https://doi.org/10.1016/0014-2999(94)90516-9

Pavone KJ et al (2016) Nitrous oxide-induced slow and delta oscillations. Clin Neurophysiol 127(1):556–564. https://doi.org/10.1016/j.clinph.2015.06.001

Peyton PJ, Wu CY (2014) Nitrous oxide-related postoperative nausea and vomiting depends on duration of exposure. Anesthesiology 120(5):1137–1145. https://doi.org/10.1097/ALN.0000000000000122

Pfenninger EG, Durieux ME, Himmelseher S (2002) Cognitive impairment after small-dose ketamine isomers in comparison to equianalgesic racemic ketamine in human volunteers. Anesthesiology 96(2):357–366. https://doi.org/10.1097/00000542-200202000-00022

Podwińska E et al (2008) Pionierzy anestezjologii--Arthur Ernest Guedel (1883-1956) [pioneers of anaesthesiology--Arthur Ernest Guedel (1883-1956)]. Anestezjol Intens Ter 40(3):192–194

Randhawa G, Bodenham A (2016) The increasing recreational use of nitrous oxide: history revisited. Br J Anaesth 116(3):321–324. https://doi.org/10.1093/bja/aev297

Sato Y et al (2005) Effect of N-methyl-D-aspartate receptor ε1 subunit gene disruption of the action of general anesthetic drugs in mice. Anesthesiology 102(3):557–561. https://doi.org/10.1097/00000542-200503000-00013

Kyoung SM et al (2020) AMPA receptor-mTORC1 signaling activation is required for neuroplastic effects of LY341495 in rat hippocampal neurons. Sci Rep 10(1):1–12. https://doi.org/10.1038/s41598-020-58017-3

Singh JB et al (2016) A double-blind, randomized, placebo-controlled, dose-frequency study of intravenous ketamine in patients with treatment-resistant depression. Am J Psychiatr 173(8):816–826. https://doi.org/10.1176/appi.ajp.2016.16010037

Swainson J et al (2019) Esketamine for treatment resistant depression. Expert Rev Neurother 19(10):899–911. https://doi.org/10.1080/14737175.2019.1640604

Todorovic SM et al (2001) Cav3.2 channel is a molecular substrate for inhibition of T-type calcium currents in rat sensory neurons by nitrous oxide. Mol Pharmacol 60(3):603–610

Trullas R, Skolnick P (1990) Functional antagonists at the NMDA receptor complex exhibit antidepressant actions. Eur J Pharmacol 185(1):1–10. https://doi.org/10.1016/0014-2999(90)90204-J

van Amsterdam J, Nabben T, van den Brink W (2015) Recreational nitrous oxide use: prevalence and risks. Regul Toxicol Pharmacol 73(3):790–796. https://doi.org/10.1016/j.yrtph.2015.10.017

Wan LB et al (2015) Ketamine safety and tolerability in clinical trials for treatment-resistant depression. J Clin Psychiatry 76(3):247–252. https://doi.org/10.4088/JCP.13m08852

Williams NR et al (2018) Attenuation of antidepressant effects of ketamine by opioid receptor antagonism. Am J Psychiatr 175(12):1205–1215. https://doi.org/10.1176/appi.ajp.2018.18020138

Wilson S (2013a) Management of child patient behavior: quality of care, fear and anxiety, and the child patient. J Endod 39(3 SUPPL):S73–S77. https://doi.org/10.1016/j.joen.2012.11.040

Wilson S (2013b) Management of child patient behavior: quality of care, fear and anxiety, and the child patient. J Endod 39(3 SUPPL):S73–S77. https://doi.org/10.1016/j.joen.2012.11.040

Wu, L. T., Pilowsky, D. J., & Schlenger, W. E. (2004). Inhalant abuse and dependence among adolescents in the United States. Journal of the American Academy of Child and Adolescent Psychiatry, 43(10), 1206–1214.. https://doi.org/10.1097/01.chi.0000134491.42807.a3

Yamakura T, Harris RA (2000) Effects of gaseous anesthetics nitrous oxide and xenon on ligand-gated ion channels: comparison with isoflurane and ethanol. Anesthesiology 93(4):1095–1101. https://doi.org/10.1097/00000542-200010000-00034

Zanos P et al (2016) NMDAR inhibition-independent antidepressant actions of ketamine metabolites. Nature 533(7604):481–486. https://doi.org/10.1038/nature17998

Zarate CA Jr et al (2006) A randomized trial of an N-methyl-D-aspartate antagonist in treatment-resistant major depression. Arch Gen Psychiatry 63(8):793–802. https://doi.org/10.1001/archpsyc.63.8.856

Zarate CA et al (2012) Replication of ketamine's antidepressant efficacy in bipolar depression: a randomized controlled add-on trial. Biol Psychiatry 71(11):939–946. https://doi.org/10.1016/j.biopsych.2011.12.010

Zorumski CF et al (2015) Treatment-resistant major depression: rationale for NMDA receptors as targets and nitrous oxide as therapy. Front Psych 6(172). https://doi.org/10.3389/fpsyt.2015.00172

[illegible]

Novel AMPA Receptor Potentiators TAK-137 and TAK-653 as Potential Rapid-Acting Antidepressants

Haruhide Kimura

Abstract Ketamine produces a rapid-onset and sustained antidepressant effect in patients with treatment-resistant depression; however, it concurrently causes undesirable psychotomimetic side effects. Accumulating evidence suggests that ketamine produces an antidepressant effect via α-amino-3-hydroxy-5-methyl-4-isoxazolepropionic acid receptor (AMPA-R) activation and consequent mammalian target of rapamycin pathway activation and brain-derived neurotrophic factor release. Thus, AMPA-R activation could be used as an alternative approach for discovering novel antidepressants with fast onset of action. However, AMPA-R potentiators carry the risk of narrow bell-shaped responses and a poor safety margin against seizure. We hypothesized that these risks are related to agonistic activity of AMPA-R potentiator and discovered two novel AMPA-R potentiators, TAK-137 and TAK-653, which have lower agonistic activity. In this chapter, I will describe our strategy for discovering AMPA-R potentiators with lower agonistic activity. I will also review the in vitro and in vivo profiles of TAK-137 and TAK-653, including their procognitive and antidepressant-like effects.

Keywords TAK-137 · TAK-653 · Ketamine · AMPA-R · Agonistic activity · Procognitive effect · Antidepressant-like effect

Abbreviations

AMPA-R	α-amino-3-hydroxy-5-methyl-4-isoxazolepropionic acid receptor
BDNF	Brain-derived neurotrophic factor
BrdU	Bromodeoxyuridine
DMTS	Delayed match-to-sample

H. Kimura (✉)
Neuroscience Drug Discovery Unit, Research, Takeda Pharmaceutical Company Limited, Fujisawa, Kanagawa, Japan
e-mail: haruhide.kimura@takeda.com

K. Hashimoto, M. Manto (eds.), *New Rapid-acting Antidepressants*, Contemporary Clinical Neuroscience, https://doi.org/10.1007/978-3-030-79790-4_5

EC_{50}	Half maximal effective concentration
ERK	Extracellular signal-regulated kinase
GABAergic	Gamma-aminobutyric acid-ergic
HNK	Hydroxynorketamine
LBD	Ligand-binding domain
MDD	Major depressive disorder
MRD	Multiple rising dose
mTOR	Mammalian target of rapamycin
NBQX	2,3-Dioxo-6-nitro-1,2,3,4-tetrahydrobenzo[f]quinoxaline-7-sulfo namide
NDI	Novelty discrimination index
NMDA-R	N-methyl-D-aspartate receptor
NORT	Novel object recognition test
NSF	Novelty-suppressed feeding
RSBM	Reduction of submissive behavior model
SPA	Scintillation proximity assay
SRD	Single rising dose
TAK-137	9-(4-Phenoxyphenyl)-3,4-dihydropyrido[2,1-c][1,2,4]thiadiazine 2,2-dioxide
TARP	Transmembrane AMPA-R regulatory proteins
TRD	Treatment-resistant depression
WKY	Wistar Kyoto

Introduction

Major depressive disorder (MDD) is one of the most prevalent psychiatric diseases and poses a considerable economic and social burden (Kessler et al. 2003). The large-scale Sequenced Treatment Alternatives to Relieve Depression (STAR*D) study showed that approximately 50% of patients with MDD responded to the first treatment; however, only 30% reached full remission. Moreover, there is a delay in the onset of action of weeks or months, even in those who responded to treatment, as well as a limited duration of efficacy (Gaynes et al. 2009). Therefore, more effective medications are required.

A single sub-anesthetic dose of ketamine (0.5 mg/kg), an N-methyl-D-aspartate receptor (NMDA-R) antagonist, produced a rapid (within 2 hours) and sustained (>1 week) antidepressant effect in patients with treatment-resistant depression (TRD) (Murrough et al. 2013; Zarate Jr. et al. 2006). However, ketamine's dissociative properties (observed at 40 minutes after ketamine infusion) and abuse potential limit its use in clinical practice (Duman et al. 2016). Thus, discovery of novel rapid-acting antidepressants with similar effects and improved safety profiles is needed.

Several hypotheses concerning the mechanisms of action underlying the antidepressant activity of ketamine have been proposed (Duman et al. 2016; Zanos and Gould 2018). Many of these hypotheses suggest the activation of α-amino-3-hydroxy-5-methyl-4-isoxazolepropionic acid receptors (AMPA-Rs) as a key downstream mechanism (Fig. 1). For example, ketamine is thought to produce paradoxical increases in glutamate release through preferential NMDA-R inhibition on gamma-aminobutyric acid-ergic (GABAergic) interneurons (disinhibition of pyramidal neurons), resulting in AMPA-R activation (Duman et al. 2016; Zanos and Gould 2018). Another hypothesis suggests that a metabolite of R-ketamine, (2R,6R)-hydroxynorketamine ((2R,6R)-HNK), may exert antidepressant-like effects through early and sustained AMPA-R activation (Duman et al. 2016; Zanos and Gould 2018) although some controversial observations on the antidepressant-like effects of (2R,6R)-HNK have been raised (Shirayama and Hashimoto 2018; Yamaguchi et al.

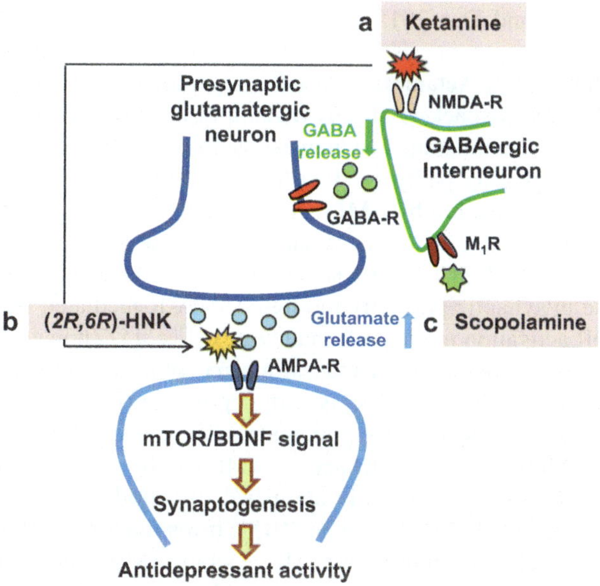

Fig. 1 Schematic diagram of the possible mechanisms of action of ketamine as an antidepressant in the prefrontal cortex. (**a**) The pyramidal neuron disinhibition hypothesis in which ketamine produces paradoxical increases in glutamate release (glutamate burst) through the preferential inhibition of N-methyl-D-aspartate receptors (NMDA-Rs) on gamma-aminobutyric acid-ergic (GABAergic) interneurons, resulting in α-amino-3-hydroxy-5-methyl-4-isoxazolepropionic acid receptor (AMPA-R) activation. AMPA-R activation leads to brain-derived neurotrophic factor (BDNF) release and mammalian target of rapamycin (mTOR) signaling activation, which, in turn, increase synaptogenesis and produce antidepressant activity. (**b**) The ketamine metabolite-mediated AMPA-R activation hypothesis in which a metabolite of R-ketamine, (2R,6R)-hydroxynorketamine ([2R,6R]-HNK), exerts antidepressant-like effects through early and sustained AMPA-R activation. (**c**) Scopolamine may also cause a glutamate burst by muscarinic M_1 receptor (M_1R) blockade on GABAergic interneurons and producing antidepressant activity via AMPA-R activation

2018). In agreement with these proposed mechanisms, the antidepressant-like effects of ketamine in animal models of depression were blocked by pretreatment with an AMPA-R antagonist, 2,3-dioxo-6-nitro-1,2,3,4-tetrahydrobenzo[f]quinoxaline-7-sulfonamide (NBQX) (Fukumoto et al. 2014; Koike et al. 2011; Maeng et al. 2008; Zhou et al. 2014).

The antidepressant-like effect of ketamine may also require the release of brain-derived neurotrophic factor (BDNF) in cortical neurons (Autry et al. 2011; Lepack et al. 2014), and activation of mammalian target of rapamycin (mTOR) via activation of Akt, and extracellular signal-regulated kinase (ERK) signaling in the prefrontal cortex (Li et al. 2010); ketamine-mediated antidepressant-like behavioral responses are blocked in *Bdnf*-null mice and by intracerebroventricular infusion of anti-BDNF antibody, an mTOR inhibitor (rapamycin), a PI3k/Akt inhibitor (LY294002), and an ERK inhibitor (U0126). Moreover, ketamine-induced mTOR activation was inhibited by the infusion of selective Akt and ERK inhibitors. BDNF promotes neuronal survival and growth and plays an important role in neuronal plasticity (Bathina and Das 2015), and mTOR regulates axon growth, dendritic arborization, and synaptogenesis, which are crucial for synaptic plasticity (Swiech et al. 2008). Importantly, ketamine-induced upregulation of BDNF and mTOR was inhibited by NBQX (Lepack et al. 2014; Zhou et al. 2014). Moreover, AMPA-R activation has been demonstrated to increase neuronal BDNF expression and activate the mTOR signaling pathway (Jourdi et al. 2009). In agreement with these findings, AMPA-R activation by AMPA-R potentiators has been shown to produce rapid antidepressant-like effects in animal models of depression (Fukumoto et al. 2014; Knapp et al. 2002). Therefore, ketamine may produce antidepressant activity via an increase in synaptic plasticity through AMPA-R-mediated BDNF upregulation and mTOR activation in the prefrontal cortex (Fig. 1) (Zhou et al. 2014).

A low dose of scopolamine, a muscarinic receptor antagonist, also produces rapid antidepressant activity in patients with depression (Ellis et al. 2014). Similar to ketamine, scopolamine induces glutamate release and AMPA-R activation in rodent models. Moreover, the blockage of AMPA-R by NBQX infusion into the rodent prefrontal cortex significantly abolished the antidepressant-like effects of scopolamine (Voleti et al. 2013; Yu et al. 2018). Investigations of receptor subtypes indicated that a selective muscarinic M_1 receptor (M_1R) antagonist produced antidepressant-like actions in rodent models (Navarria et al. 2015; Witkin et al. 2014). Thus, AMPA-R activation may also play a critical role in the rapid-acting antidepressant-like effects of scopolamine (Duman et al. 2016).

In addition to ketamine and scopolamine, different antidepressant classes including selective serotonin-reuptake inhibitors, dual- and triple-reuptake inhibitors, dopaminergic agents, tricyclics, and electroconvulsive therapy induce BDNF increase. Moreover, the behavioral responses of these agents are canceled in mice with reduced BDNF expression (Chen et al. 2006; Jiang et al. 2014; Yu et al. 2012). Thus, upregulation of BDNF signaling pathways seems to play a critical role in producing antidepressant activity. However, AMPA-R potentiators called low impact ampakines such as CX516, CX717, Org24448 (CX691), and CX1739 may not induce BDNF production in neurons (Carmichael 2012). In cultured rat primary neurons under our experimental conditions, CX516 induced a small AMPA-R current enhancement but did not induce BDNF production at up to 2000μM (Suzuki et al. 2019a). Therefore, an AMPA-R potentiator with the potential to induce BDNF release could be a novel therapy for the treatment of MDD, including TRD.

Establishment of a Novel Drug Screening Strategy for AMPA-R Potentiators

AMPA-Rs belong to a family of ionotropic glutamate receptors that mediates the majority of fast synaptic transmission at excitatory synapses and synaptic plasticity in the central nervous system, which is essential for learning and memory (Derkach et al. 2007; Kew and Kemp 2005; Lynch 2004). AMPA-R potentiators have improved cognitive performance in multiple tasks (Lynch 2004; Tanaka et al. 2019). AMPA-Rs are composed of four subunits (GluA1–4), which have flip (i) and flop (o) splice isoforms. The flip and flop splice variants affect the rate and extent of heteromeric AMPA-R desensitization and their sensitivity to AMPA-R potentiators (Dingledine et al. 1999; Kew and Kemp 2005). AMPA-Rs are most likely tetramers generated by the assembly of one or more of these subunits (Greger et al. 2017). RNA editing of a GluA2 Q/R site affects the permeability of divalent cations, including Ca^{2+}. Thus, AMPA-Rs in the brain containing GluA2 are considered Ca^{2+} impermeable (Dingledine et al. 1999; Seeburg and Hartner 2003). Differential assembly with auxiliary accessory transmembrane proteins, such as transmembrane AMPA-R regulatory proteins (TARPs), regulates AMPA-R trafficking, kinetics, and synaptic waveforms (Greger et al. 2017; Kato et al. 2010). The extracellular portion of each AMPA-R subunit consists of two domains: a ligand-binding domain (LBD) and an amino-terminal domain (ATD), which is also referred to as the N-terminal domain (Greger et al. 2017). Thus far, no ligands have been shown to influence AMPA-R biophysical properties by binding to the ATD (Fleming and England 2010).

AMPA-R agonists, which can activate all AMPA-Rs, including resting receptors, have been shown to induce AMPA-R desensitization and seizures (Beattie et al. 2000; Ehlers 2000; Yamada 1998). Thus, AMPA-R potentiators are considered to have better pharmacological profiles because of their selective enhancement of physiological AMPA-R activation in the brain. However, AMPA-R potentiators, such as LY451646 (an active isomer of LY404187), have been reported to elicit bell-shaped dose responses to various pharmacological effects. For example, at 0.5 mg/kg in rats, LY404187 significantly increased glucose utilization and c-fos induction in certain brain regions, such as the parietal cortex, whereas it showed little or no activity at 0.05 or 5 mg/kg (Fowler et al. 2004). Subchronic LY451646 treatment (7 days) induced significant hippocampal BDNF production at 0.5 mg/kg, but not at ≤0.25 or 1.0 mg/kg in rats (Mackowiak et al. 2002). Acute LY451646 administration at 0.05 mg/kg, but not at 0.025 or 0.125 mg/kg, induced a significant increase in the number of clusters of neuronal progenitor cells in the rat dentate gyrus (Bai et al. 2003). At the 2009 Society for Neuroscience meeting, a bell-shaped response to LY451646 in the radial arm maze task was demonstrated (Majchrzak et al. 2009) with LY451646 at 0.1 mg/kg, but not at ≤0.032 or ≥0.32 mg/kg, significantly recovering MK-801-induced deficits in working memory. Moreover, some AMPA-R potentiators have a relatively narrow margin of safety between cognitive enhancement and seizure induction. For example, in rats, the exposure margin of LY451646 between memory enhancement in the novel object recognition test (NORT) and lack of seizures assessed using area under the curve for brain drug concentration-time (AUC_{brain}) values and brain maximum concentration (C_{max}) values was only 3.1- and 7.5-fold, respectively (Kunugi et al. 2019). Therefore, we decided to search for novel AMPA-R potentiators with lower risks of these issues.

Multiple approaches have been applied to discovering AMPA-R potentiators with improved efficacy and safety profiles. For example, AMPA-R potentiators with subunit preference, flip or flop splice isoform preference, desensitization or deactivation selective modulation, and TARP family selective modulation have been considered (Alt et al. 2006; Fleming and England 2010; Kew and Kemp 2005; Lynch 2004). However, to our knowledge, ideal BDNF-inducible AMPA-R potentiators with low bell-shaped dose response and seizure risks have not been discovered.

We explored in vitro assays that detected bell-shaped dose responses to discover AMPA-R potentiators with lower bell-shaped response risks. We found that LY451646 (Fig. 2a) and LY451395 started to exhibit a bell-shaped response at concentrations in which they began to show agonistic effects on BDNF protein production in primary neurons (Fig. 2b) (Kunugi et al. 2018). Potent agonistic effects of LY451646 and LY451395 were also confirmed by a patch-clamp study and Ca^{2+} influx assay (Fig. 2c, d). In primary neurons, LY451646 and LY451395 induced AMPA-R-mediated currents or Ca^{2+} influx in the presence and absence of agonists (AMPA) (Kunugi et al. 2018; Kunugi et al. 2019). We hypothesized that the agonistic property was related to bell-shaped responses in BDNF production and endeavored to discover novel AMPA-R potentiators with lower agonistic effects. Unfortunately, patch-clamp studies and BDNF assays are labor-intensive and not suitable for high-throughput screening. Thus, Ca^{2+} influx was considered as an alternative method. However, to our surprise, both LY451646 and LY451395 induced Ca^{2+} influx in the presence and absence of AMPA in primary neurons, whereas they only induced Ca^{2+} influx in the presence of glutamate in GluA1i and TARPs γ2 expressing CHO cells (GluA1i CHO cells) (Fig. 2d, e) (Kunugi et al. 2018; Kunugi et al. 2019).

We searched for novel AMPA-R potentiators with low agonistic activity using Ca^{2+} influx in primary neurons and discovered two molecules: 2-((5-methyl-3-(trifluoromethyl) -1H-pyrazol-1-yl)acetyl)amino)-4,5,6,7-tetrahydro-1-benzothiophene-3-carboxamide (HBT1) and (3S)-1-(4-tert-butylphenyl)-N-((1R)-2-(dimethylamino)-1-phenylethyl)-3-isobutyl-2-oxopyrrolidine –3-carboxamide (OXP1) (Fig. 3a, b) (Kunugi et al. 2018). In a patch-clamp study using primary neurons, HBT1 induced

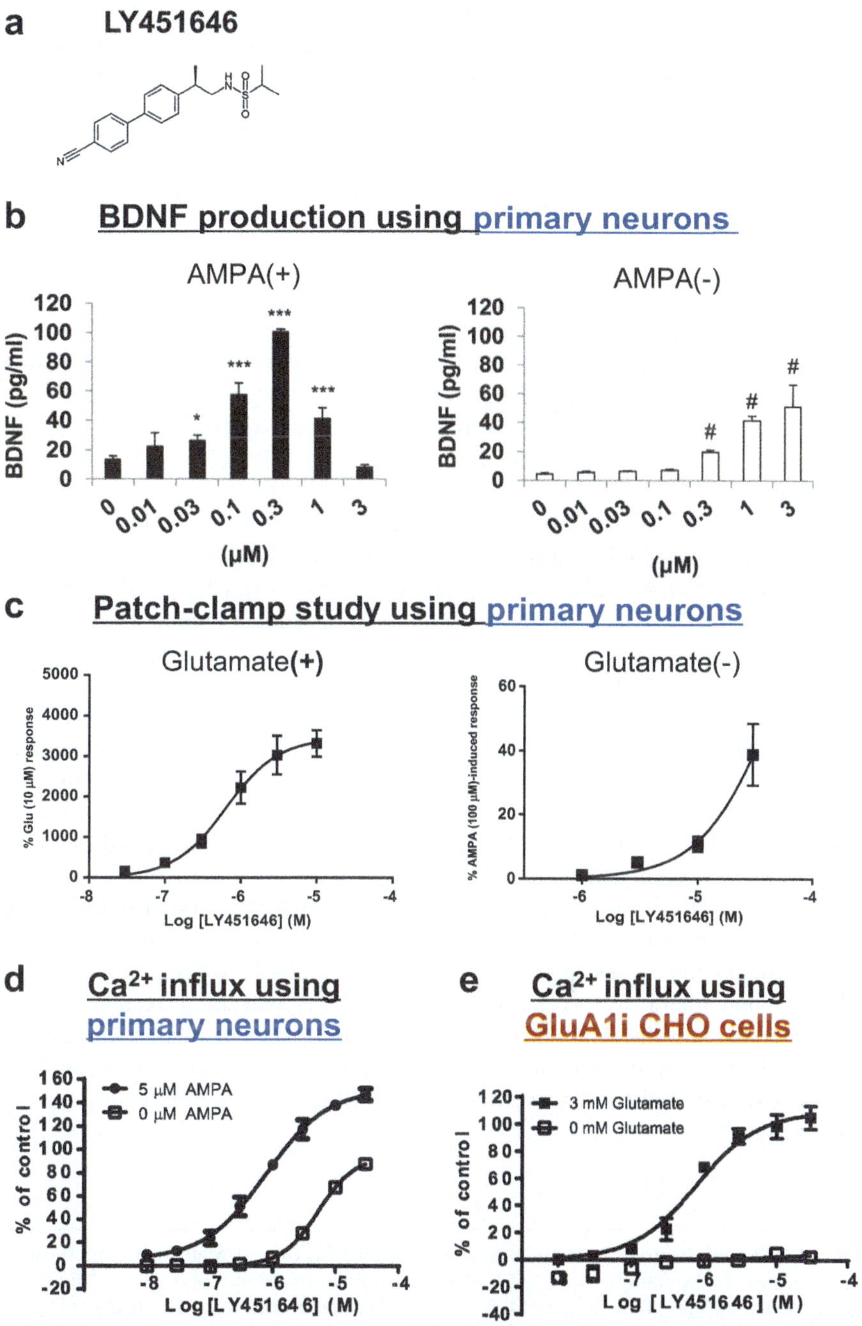

Fig. 2 LY451646 showed a bell-shaped response for brain-derived neurotrophic factor (BDNF) production at concentrations at which it started to induce agonistic activity in primary neurons. (**a**) Chemical structure of LY451646. (**b**) Effect of LY451646 on BDNF protein levels in primary

AMPA-R-mediated currents in the presence of AMPA only, whereas OXP1 induced AMPA-R-mediated currents in the presence and absence of AMPA (Fig. 3c) (Kunugi et al. 2018). Moreover, HBT1 concentration-dependently increased BDNF protein levels in the presence of AMPA with no bell-shaped dose response, and HBT1 at up to 3μM showed no agonistic effect on BDNF protein induction. In contrast, OXP1 started to exhibit a bell-shaped response where it began to show an agonistic effect on BDNF production in primary neurons (Fig. 3d) (Kunugi et al. 2018). AMPA-R potentiator subunit selectivity may affect their functional outcomes (Ward et al. 2010). However, HBT1 and OXP1 showed little subunit selectivity in Ca^{2+} influx assays using CHO cells expressing GluA1–4i and TARPs γ2 or GluA1–4o and TARPs γ2 (Kunugi et al. 2018). HBT1 required an approximately tenfold higher concentration of glutamate than OXP1 to activate AMPA-Rs, suggesting their different mechanisms of action (Kunugi et al. 2018).

We examined the binding properties of HBT1 and OXP1 to AMPA-Rs. The hydrophilic pocket at the dimer interface of the AMPA-R LBD is a binding pocket for several AMPA-R potentiators (Kaae et al. 2007; Sobolevsky et al. 2009). We investigated the interaction between HBT1 and the LBD using a scintillation proximity assay (SPA), which measured the binding of radiolabeled HBT1 ([^{3}H]-HBT1) to a His-tagged GluA2o LBD protein (His-LBD) (Fig. 4a). [^{3}H]-HBT1 specifically bound to the LBDs on AMPA-Rs in a glutamate-dependent manner (Fig. 4b), with Kd = 543 nM in the presence of 100μM of glutamate. Similarly, glutamate-dependent binding of [^{3}H]-HBT1 to rat hippocampal membranes with Kd = 416 nM was observed. Crystal structures of the GluA2o LBD/glutamate/HBT1 complex revealed that HBT1 bound to the cleft formed by the dimer interface between two protomers with a compound-to-protomer stoichiometry of 1:2 to stabilize the dimer interface in the active conformation. In contrast, a specific interaction between the His-LBD and [^{3}H]-OXP1 was not detected. Preliminary data suggested that OXP1 binds to the ATDs of AMPA-Rs. To our surprise, in the absence of agonists, co-stimulation by HBT1 and OXP1 robustly activated AMPA-Rs on primary neurons, but not those on GluA1i CHO cells (Kunugi et al. 2018). Functional assays using non-neuronal cells expressing GluA1i or other subunits have been widely used for AMPA-R potentiator discovery (Ward et al. 2010). However, our results suggested that understanding the relationship between binding sites or modes and the functional properties of each compound using both recombinant and physiological receptors was critical for discovering novel AMPA-R potentiators with lower agonistic activity.

Fig. 2 (continued) neurons. Data are presented as mean ± SD (n = 3). Statistically significant differences between the controls and LY451646 in the presence or absence of α-amino-3-hydroxy-5-methyl-4-isoxazolepropionic acid (AMPA) (1μM) were determined using Dunnett's ($*P \leq 0.05$, $***P \leq 0.001$) and one-tailed Williams' ($\#P \leq 0.025$) tests, respectively. (**c**) Effects of LY451646 on AMPA-R-mediated currents on the presence or absence of glutamate (10μM) in an electrophysiology study using primary neurons. Data are presented as mean ± SEM (n = 4–9). (**d**, **e**) Effect of LY451646 on Ca^{2+} influx in (**d**) primary neurons and (**e**) GluA1i CHO cells in the presence or absence of an agonist. Data are presented as mean ± SD (n = 3)

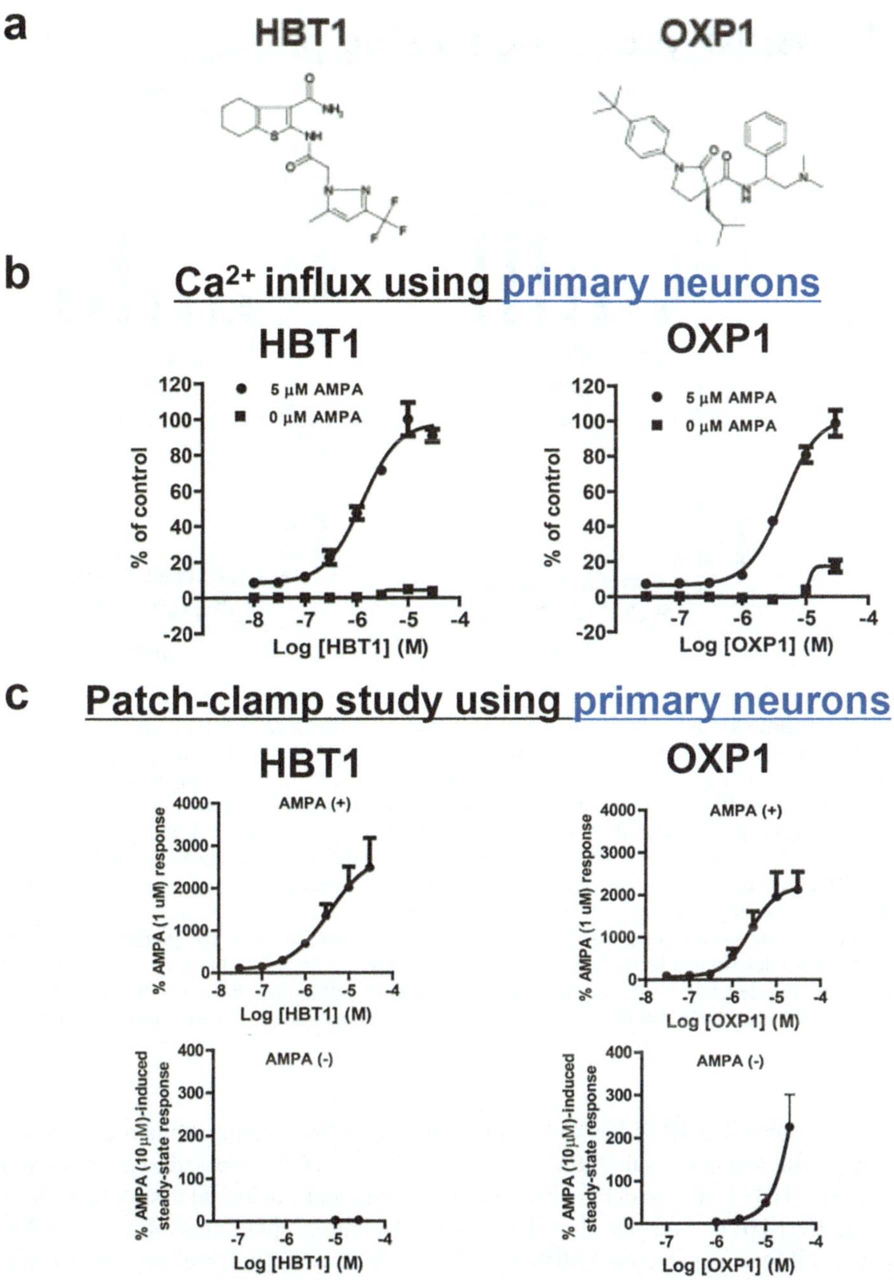

Fig. 3 Differential agonistic effects of HTB1 and OXP1 on Ca^{2+} influx, α-amino-3-hydroxy-5--methyl-4-isoxazolepropionic acid receptor (AMPA-R)-mediated currents, and brain-derived neurotrophic factor (BDNF) production and their risks of bell-shaped dose responses for BDNF production in the presence of a low concentration of AMPA (1μM) in primary neurons. (**a**)

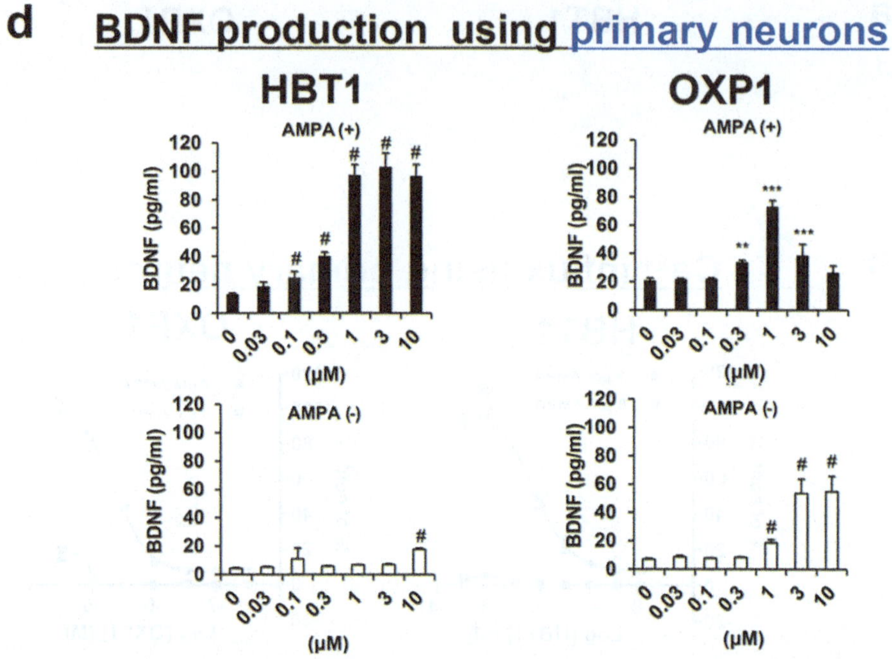

Fig. 3 (continued) Chemical structure of HBT1 and OXP1. (**b**) Effect of HBT1 and OXP1 on Ca^{2+} influx in primary neurons in the presence or absence of AMPA (5μM). Data are presented as mean ± SD ($n = 3$). (**c**) Effects of HBT1 and OXP1 on AMPA-R-mediated currents in the presence or absence of AMPA (1μM) in a patch-clamp study using primary neurons. Data are presented as mean ± SEM ($n = 4$–9). (**d**) Effect of HBT1 and OXP1 on BDNF protein levels in primary neurons. Data are presented as mean ± SD ($n = 3$). Statistically significant differences between the controls and HBT1 in the presence and absence of AMPA (1μM) were determined using the one-tailed Williams' test (#$P \leq 0.025$)
Statistically significant differences between the controls and OXP1 in the presence of AMPA (1μM) were determined using Dunnett's test (**$P \leq 0.01$, ***$P \leq 0.001$) because of the lack of concentration-dependent effects. Statistically significant differences between the controls and OPX1 in the absence of AMPA were determined using the one-tailed Williams' test (#$P \leq 0.025$)

Unfortunately, HBT1 has a poor absorption, distribution, metabolism, and excretion profile and is not suitable for in vivo studies. Thus, we decided to discover and optimize HBT1-site binders using various binding and functional assays (e.g., SPA using [^{3}H]-HBT1 and the His-LBD, X-ray crystallographic analysis, Ca^{2+} influx using cell lines expressing AMPA-R and primary neurons, patch-clamp study using primary neurons, and BDNF production assay using primary neurons).

Characterization of AMPA-R Potentiators Containing a Dihydropyridothiadiazine 2,2-Dioxide Skeleton with a Different Agonistic Effect

We screened a chemical library by SPA using [^{3}H]-HBT1 and the His-LBD (Fig. 4a) and discovered a novel chemotype (dihydropyridothiadiazine 2,2-dioxides). In the SPA, two dihydropyridothiadiazine 2,2-dioxide derivatives, Compound-1 and Compound-2 (Fig. 5a), inhibited binding between [^{3}H]-HBT1 and the LBD with an inhibition constant (Ki) of 0.082 and 0.019μM, respectively (Kunugi et al. 2019). Interestingly, in the Ca^{2+} influx assay using primary neurons, Compound-1 may have a lower agonistic effect than that of Compound-2. The half maximal effective concentration (EC_{50})/maximum effect (E_{max}) of Compound-1 in the presence and absence of AMPA was 0.43μM /124% and 2.9μM/21%, respectively, whereas that of Compound-2 in the presence and absence of AMPA was 0.13μM/145% and 1.2μM/96%, respectively (Fig. 5b). Consistent with these observations, [^{3}H]-Compound-1 bound to hippocampal membranes in the presence of 3 mM of glutamate, but not in the absence of glutamate, whereas [^{3}H]-Compound-2 bound to hippocampal membranes in the presence and absence of glutamate (Kunugi et al. 2019).

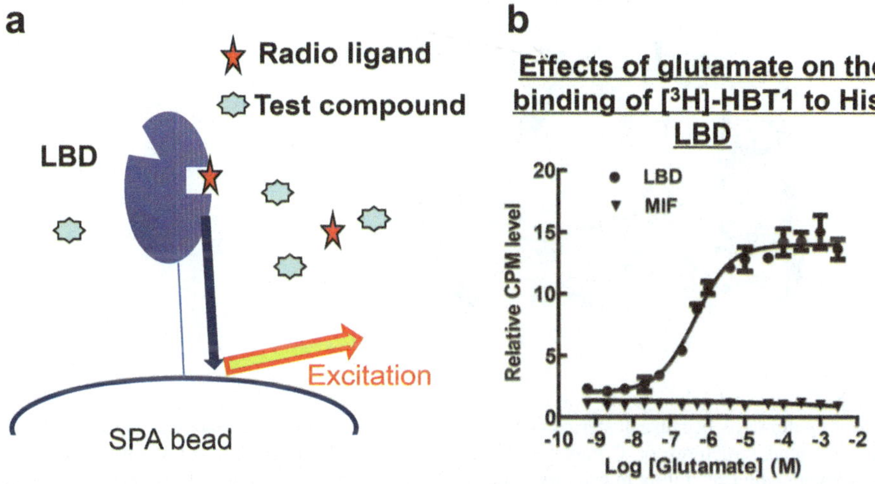

Fig. 4 Establishment of a high-throughput binding assay measuring the binding of tritium-labeled HBT-1 ([^{3}H]-HBT1) to a histidine-tagged GluA2o ligand-binding domain protein (His-LBD) using a scintillation proximity assay (SPA). (**a**) Schematic diagram of the binding assay using [^{3}H]-HBT1 and His-LBD. The His-LBD was fixed to the SPA beads, and the binding of [^{3}H]-HBT1 to the His-LBD was measured using a microplate scintillation counter. Displacement studies were conducted to discover compounds with a binding affinity to the HBT1 binding site on the LBD. (**b**) [^{3}H]-HBT1 bound to the His-LBD in a glutamate-dependent manner. His-tagged purified recombinant macrophage migration inhibitory factor (MIF) was used as a control protein. Nonspecific MIF binding was not observed. Data are presented as mean ± SD (n = 3–4)

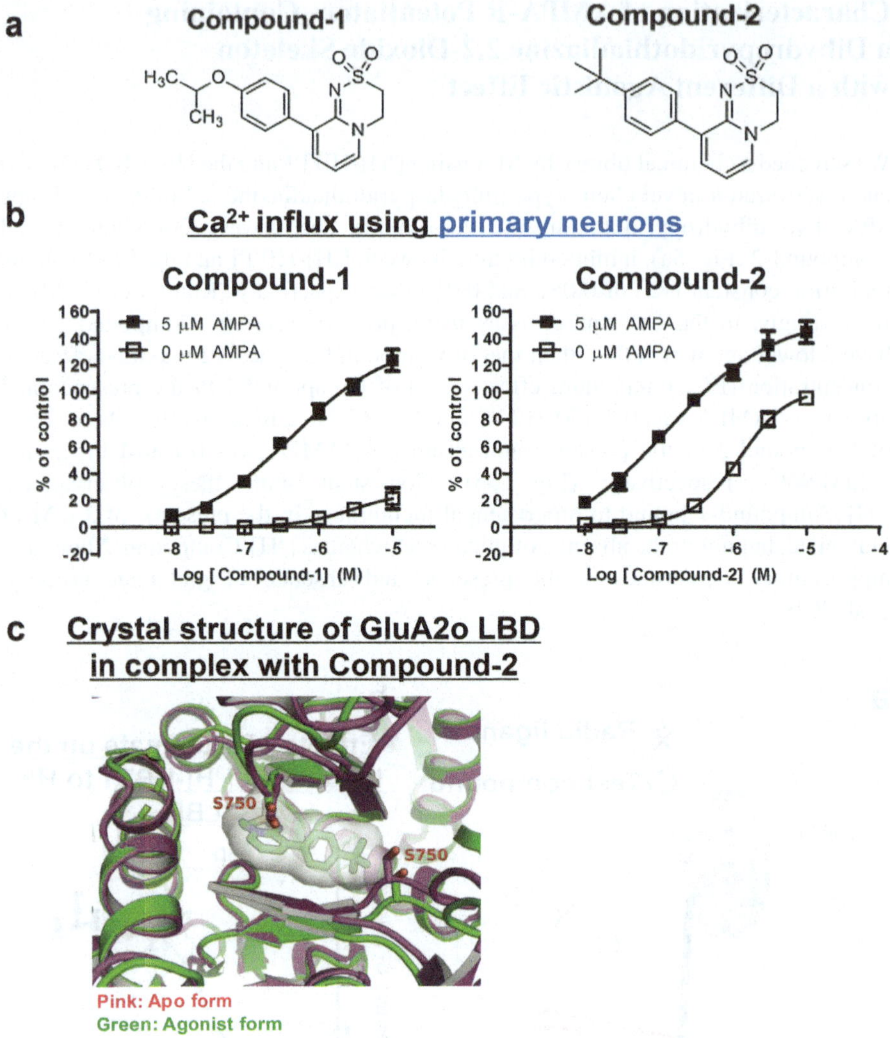

Fig. 5 Compound-1 had lower agonistic activity than Compound-2 on Ca^{2+} influx in primary neurons potentially because of steric interference between the bulky isopropoxy group and Ser750. (**a**) Chemical structures of Compound-1 and Compound-2. (**b**) Effects of Compound-1 and Compound-2 on Ca^{2+} influx in primary neurons in the presence or absence of α-amino-3-hydroxy−5-methyl-4-isoxazolepropionic acid (AMPA) (5μM). Data are presented as mean ± SD (n = 3). (**c**) X-ray crystallographic analysis. Main-chain atoms around Ser750 located close to the potentiator-binding cleft were turned aside by agonist binding (from pink to green). Compound-1, but not Compound-2, could not bind to GluA2o in the absence of glutamate (apo form) because of steric interference between the bulky isopropoxy group and Ser750

X-ray crystallographic analysis of Compound-1 and Compound-2 interactions with the LBD revealed that both compounds bound to the cleft composed of the dimer interface between two protomers. Interestingly, structural comparison of the full-length GluA2o between the channel-open and channel-closed states showed that the main-chain atoms around Ser750, which were closely located to the potentiator-binding cleft, were turned aside by agonist binding (Fig. 5c). Moreover, the terminal substituent of Compound-2 (tert-butyl group), which is different from that of Compound-1 (isopropoxy group), was located in the vicinity of the serine residue (Fig. 5c). From these data, we speculated that Ser750 may prevent Compound-1, but not Compound-2, from binding in the "apo" agonist-free state because of steric interference.

Ser750 in GluA2o corresponds to Ser743 in GluA1i. Thus, we introduced a mutation at Ser743 (S743A or S743V) in GluA1i to modify the possible steric interference between Compound-1 or Compound-2 and AMPA-R. Theoretically, S743A lowers this steric interference, whereas S743V increases it. Interestingly, with a Ca^{2+} influx assay using CHO cells at lower concentrations (0.3 and 1μM) of glutamate, the maximum responses of Compound-2 to the activation of WT GluA1i were higher compared with Compound-1. S743A increased the maximum responses of Compound-1, but not those of Compound-2, at lower concentrations of glutamate. Moreover, S743V lowered the maximum responses of Compound-2 at lower concentrations of glutamate (Kunugi et al. 2019). Thus, in the presence of suboptimal glutamate concentrations, S743 interactions may, at least in part, relate to the maximum responses of these compounds.

Discovery and Characterization of TAK-137

We designed dihydropyridothiadiazine 2,2-dioxides with minimum binding affinity to the LBD in the channel-closed state with steric interference at Ser743 in GluA1i (Ser750 in GluA2o) and discovered 9-(4-Phenoxyphenyl)-3,4-dihydropyrido[2,1-c][1,2,4]thiadiazine 2,2-dioxide (TAK-137), which is a novel AMPA-R potentiator with lower agonistic effects (Fig. 6a) (Kunugi et al. 2019).

In Vitro Characterization of TAK-137 Using LY451646 as a Control

TAK-137 bound to the cleft composed of the dimer interface between two protomers, and a bulky phenoxy group of TAK-137 was located in the vicinity of Ser750 (Fig. 6b). [^{3}H]-TAK-137 bound to the His-LBD and hippocampal membranes in a glutamate-dependent manner. TAK-137 and LY451646 inhibited

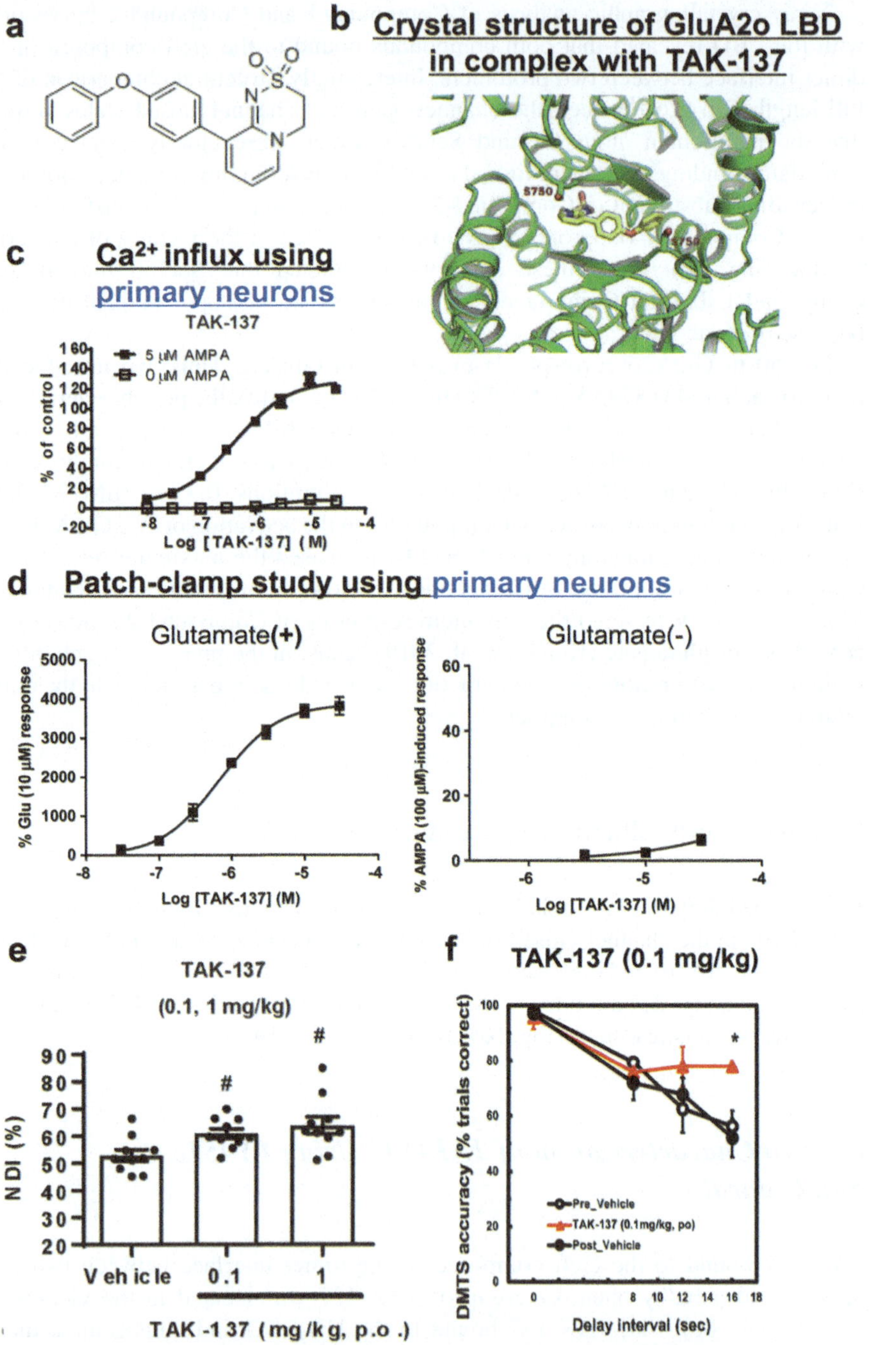

Fig. 6 TAK-137 showed lower agonistic activity on Ca^{2+} influx and α-amino-3-hydroxy-5-methyl-4-isoxazolepropionic acid receptor (AMPA-R)-mediated currents in primary neurons and

binding between [^{3}H]-HBT1 and the hippocampal membranes with Kis of 0.025 and 0.031μM, respectively. TAK-137 and LY451646 induced Ca^{2+} influx in an agonist-dependent manner in GluA1i CHO cells (EC_{50}, 1.0 and 0.77μM, respectively). TAK-137 induced Ca^{2+} influx in an agonist-dependent manner in primary neurons, whereas LY451646 induced Ca^{2+} influx in the presence and absence of AMPA (Figs. 2d and 6c). The EC_{50}/E_{max} of TAK-137 in the presence and absence of AMPA was 0.42μM/120% and 2.4μM/7.6%, respectively, and that of LY451646 was 0.78μM/147% and 5.4μM/87.6%, respectively. In agreement with the Ca^{2+} influx assay, TAK-137 induced AMPA-R-mediated currents in the presence of glutamate only, whereas LY451646 induced AMPA-R-mediated currents in the presence and absence of glutamate in an electrophysiological study using primary neurons (Figs. 2c and 6d). Thus, TAK-137 may have a lower agonistic effect than LY451646. A slow recovery from potentiation may cause sustained AMPA-R activation and induce seizures. However, in an electrophysiological study using primary neurons, AMPA-R activation induced by TAK-137 or LY451646 showed similar recovery durations (90 seconds) from potentiation. TAK-137 and LY451646 showed minimal subunit selectivity (Kunugi et al. 2019). TAK-137 and LY451646 similarly increased the potency and binding affinity of AMPA for AMPA-Rs (Suzuki et al. 2019a). Overall, the profiles of TAK-137 and LY451646, excluding their agonistic activities, were very similar.

Potent Procognitive Effects and Lower Bell-Shaped Response and Seizure Risks with TAK-137

We examined the effects of TAK-137 and LY451646 on cognitive function using the NORT in rats. TAK-137 at 0.1 and 1 mg/kg, p.o., but not at ≤0.03 mg/kg, p.o., significantly increased the novelty discrimination index (NDI) in rat NORTs (Fig. 6e). LY451646 at 1 and 3 mg/kg, p.o., but not at 0.3 mg/kg, p.o., significantly increased the NDI. Seizure induction after acute treatment with these compounds revealed that TAK-137 had a 116-fold (AUC_{brain}) and 43.7-fold (brain C_{max}) exposure margin in rats, whereas LY451646 only had a 3.1-fold (AUC_{brain}) and 7.5-fold (brain C_{max})

Fig. 6 (continued) improved cognitive functions over a broad dosage range in rats and monkeys. (**a**) Chemical structure of TAK-137. (**b**) Crystal structure of the GluA2o LBD in complex with TAK-137. Steric interference between the bulky phenoxy group and Ser750 can be expected in the absence of glutamate. (**c**) Effect of TAK-137 on Ca^{2+} influx in primary neurons in the presence or absence of AMPA (5μM). Data are presented as mean ± SD (n = 3). (**d**) Effects of TAK-137 on AMPA-R-mediated currents in the presence or absence of glutamate (10μM) in a patch-clamp study using primary neurons. Data are presented as mean ± SEM (n = 4–9). (**e**) Cognition-enhancing effects of TAK-137 in novel object recognition test (NORT) using rats. Data are presented as mean ± SEM (n = 10). Significant differences from the vehicle-treated group were analyzed using the one-tailed Williams' test ($^{\#}P \leq 0.025$). (**f**) Cognition-enhancing effects of TAK-137 in delayed match-to-sample (DMTS) task using monkeys. Data are presented as mean ± SEM (n = 4). Significant differences from the vehicle-treated group were analyzed using two-way analysis of variance (ANOVA) ($^{*}P \leq 0.05$)

exposure margin. Thus, TAK-137 may have a wider exposure margin between cognitive improvement and seizure induction than LY451646 in rats (Kunugi et al. 2019).

In monkeys, TAK-137 at 0.03, 0.1, and 1 mg/kg, p.o., but not at 0.01 mg/kg, p.o., significantly increased delayed match-to-sample (DMTS) accuracy in the DMTS task (effective at a ≥30-fold dose range) (Fig. 6f). Moreover, LY451646 at 0.1 mg/kg, s.c., but not at 0.03 mg/kg, s.c., significantly increased DMTS accuracy, and LY451646 at 1 mg/kg, s.c., induced vomiting (effective at a <ten-fold dose range). In monkeys, TAK-137 may show greater efficacy for cognitive improvement over a broader dosage range than LY451646 (Kunugi et al. 2019).

The effects of TAK-137 on social functioning and cognition were further characterized using a cognitive battery. TAK-137 significantly ameliorated MK-801-induced social interaction deficits in rats. TAK-137 also enhanced attention in the five-choice serial reaction time task. Moreover, TAK-137 improved MK-801-nduced working memory deficits in the radial arm maze test in rats as well as ketamine-induced deficits in a DMTS task in monkeys. TAK-137 improved the cognitive flexibility of subchronic phencyclidine-treated rats in reversal learning (Tanaka et al. 2019).

Antidepressant-Like Activity of TAK-137 in Rodents

In the presence of a low concentration (3μM) of AMPA, TAK-137 at 0.03–0.3μM increased the levels of the phosphorylated form of Akt, ERK, mTOR, and p70S6 kinase (p70S6K) without influencing their total protein levels in rat primary cortical neurons. This finding suggests that TAK-137 induced the activation of the mTOR signaling pathway. In rat primary cortical neurons, TAK-137 at ≥0.1μM in the presence of a low concentration (1μM) of AMPA robustly increased BDNF protein levels (Suzuki et al., 2019a).

In patients with mood disorders, recovery from reduced hippocampal neurogenesis may be implicated in the antidepressant effects of various drugs and treatments (Lucassen et al. 2010). Ketamine may accelerate doublecortin-positive adult hippocampal neural progenitor differentiation into functionally mature neurons via the activation of the TrkB-dependent ERK pathway, producing sustained antidepressant effects (Ma et al. 2017). Interestingly, repeated administration of TAK-137, but not LY451646, for 4 days increased the number of bromodeoxyuridine (BrdU)-positive cells (progenitor cell proliferation) in mouse or rat hippocampi over a broad dosage range (at least 0.3–3 mg/kg, p.o., in rats) (Fig. 7a).

Current antidepressants take several weeks to achieve their therapeutic effects; however, after acute treatment, these drugs show efficacy in some behavioral tasks, such as the forced swim and tail suspension tests. Ketamine induces hyperlocomotion following acute treatment, which prevents the accurate assessment of antidepressant-like effects using preclinical behavior studies. Thus, we selected a paradigm in which current antidepressants take a few weeks to achieve their therapeutic effects and that could detect ketamine's efficacy after its plasma clearance

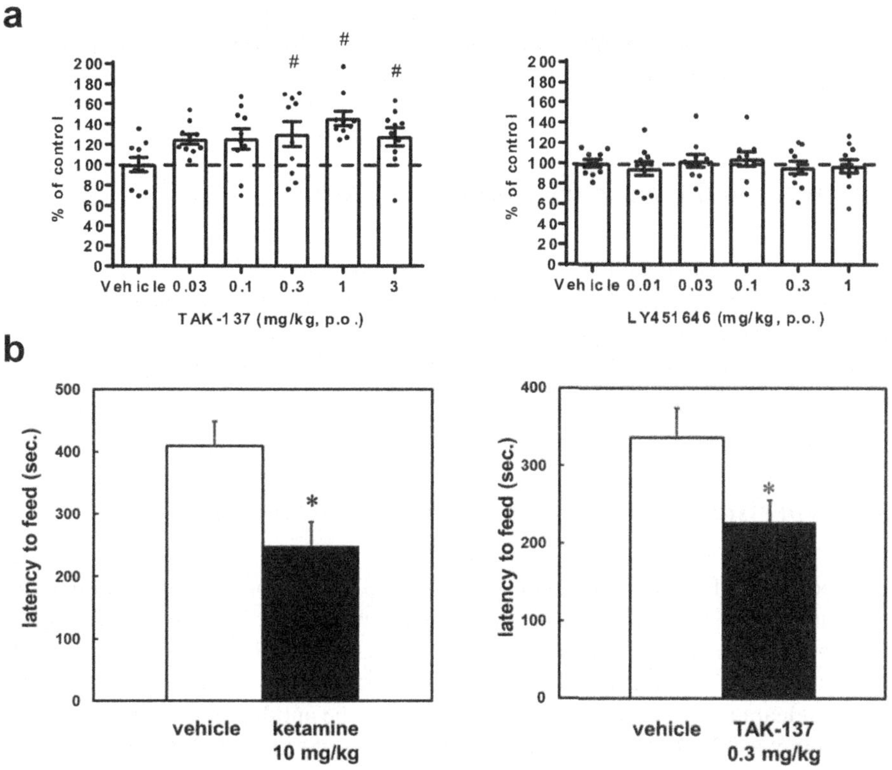

Fig. 7 TAK-137 promoted hippocampal neuronal progenitor cell proliferation and produced antidepressant-like activity in rats. (**a**) Effects of TAK-137 and LY451646 on the number of bromodeoxyuridine (BrdU)-positive cells in the rat hippocampus. TAK-137 or LY451646 was orally administered to rats for 4 d, and then BrdU was administrated. Tissues were isolated on the next day. Data are presented as mean ± SEM ($n = 9–10$). Significant differences from the vehicle-treated group were analyzed using the one-tailed Williams' test ($^{\#}P \leq 0.025$). (**b**) A novelty-suppressed feeding test (NSF test) using food-deprived Wistar Kyoto (WKY) rats was conducted 1 d after administering a single ketamine dose (10 mg/kg, i.p., $n = 6$) or TAK-137 (0.3 mg/kg, p.o., $n = 9$) for 3 d consecutively. Data are presented as mean ± SEM

(1 day after administration when it no longer induces hyperlocomotion). Chronic administration of current antidepressants and single treatment with ketamine can produce antidepressant-like activity in the novelty-suppressed feeding (NSF) test (Li et al. 2010; Marcussen et al. 2008). Therefore, the NSF test was used to characterize the antidepressant-like activity of TAK-137.

The antidepressant-like effects of ketamine and TAK-137 were evaluated in food-deprived Wistar Kyoto (WKY) rats using the NSF test. With the NSF test, antidepressant-like effects are evaluated by measuring the latency to eat a food pellet in a novel environment (Marcussen et al. 2008). Ketamine at 10 mg/kg, i.p., significantly reduced the latency to feed at 1 day after administration (Fig. 7b). Under these conditions, repeated pretreatment with TAK-137 for 3 days significantly shortened the latency to feed at 1 day after the final dose (Fig. 7b) (Suzuki et al. 2019b). In WKY rats performing the NSF test, TAK-137 may exert antidepressant-like effects.

Discovery and Characterization of TAK-653

TAK-653 is a potent and selective AMPA-R potentiator with minimal agonistic effects. TAK-653 induced Ca^{2+} influx in an agonist-dependent manner in GluA1i CHO cells and primary hippocampal neurons. Similar to TAK-137, TAK-653 dose-dependently activated mTOR signaling and induced BDNF protein production in rat primary cortical neurons (Suzuki et al. 2018, 2021).

Antidepressant-like Activity of TAK-653 in Rodents

The reduction of submissive behavior model (RSBM) is another paradigm that met our criteria. Current antidepressants take several weeks (3–5 weeks) to show significant antidepressant-like effects (Malatynska et al. 2002; Pinhasov et al. 2005). In the rat RSBM, we detected the efficacy of ketamine at 30 mg/kg, i.p., after its plasma clearance (at 1 day after administration). The antidepressant-like activity of ketamine was blocked by pretreatment with 10 mg/kg, i.p., of NBQX, indicating that the antidepressant-like effects of ketamine are likely related to AMPA-R activation in the RSBM. Consistent with this finding, repeated treatment with TAK-653 produced significant antidepressant-like effects with a potentially fast onset. Unlike ketamine, TAK-653 did not induce hyperlocomotion in rats, which is one of the behavioral changes linked to psychotomimetic side effects in humans (Suzuki et al. 2018). These results suggest that TAK-653 may be a promising drug for the treatment of MDD, including TRD, with a potentially superior safety profile.

Pharmacokinetic Properties of TAK-653 After Single and Multiple Rising Doses in Healthy Volunteers

The safety and tolerability of TAK-653 were tested in healthy volunteers in single rising dose (SRD) and multiple rising dose (MRD) settings. TAK-653 was well tolerated in healthy subjects in the SRD and MRD cohorts at all doses tested. TAK-653 exposure was approximately dose proportional. Because of its long half-life, three- to fourfold TAK-653 accumulation was observed after once per day administration. Comparable TAK-653 concentrations between the cerebrospinal fluid and unbound plasma are suggestive of its rapid brain penetration (Asgharnejad et al. 2018).

Conclusions

The discovery of the rapid antidepressant actions of ketamine represents a major advance in the treatment of mood disorders. However, ketamine's potential for abuse and psychotomimetic side effects poses challenges concerning its use in conventional clinical settings. Recent progress in understanding the mechanisms of action of ketamine, as well as those of scopolamine, has provided a strong rationale for developing AMPA-R potentiators as novel fast-acting antidepressants with improved safety profiles. However, AMPA-R potentiators have issues, such as lacking BDNF production potential, narrow bell-shaped dose responses, and poor safety margins against seizure. In rodents, the novel BDNF-inducible AMPA-R potentiators with lower agonistic activity, TAK-137 and TAK-653, showed potent procognitive and antidepressant-like activities over broad dosage ranges without inducing the hyperlocomotion associated with psychotomimetic side effects. The clinical characterization of TAK-653 is ongoing.

Acknowledgments I thank Atsushi Suzuki, Maiko Tanaka, and Akiyoshi Kunugi for their helpful comments regarding the content and organization of this manuscript.

References

Alt A, Nisenbaum ES, Bleakman D, Witkin JM (2006) A role for AMPA receptors in mood disorders. Biochem Pharmacol 71(9):1273–1288

Asgharnejad M, Zicha S, Xu L et al (2018) Pharmacokinetic and pharmacodynamic properties of the investigational AMPA receptor–positive allosteric modulator TAK-653 after single and multiple rising doses in healthy volunteers. Neuropsychopharmacology 43(suppl):M148

Autry AE, Adachi M, Nosyreva E et al (2011) NMDA receptor blockade at rest triggers rapid behavioural antidepressant responses. Nature 475(7354):91–95

Bai F, Bergeron M, Nelson DL (2003) Chronic AMPA receptor potentiator (LY451646) treatment increases cell proliferation in adult rat hippocampus. Neuropharmacology 44(8):1013–1021

Bathina S, Das UN (2015) Brain-derived neurotrophic factor and its clinical implications. Arch Med Sci 11(6):1164–1178

Beattie EC, Carroll RC, Yu X et al (2000) Regulation of AMPA receptor endocytosis by a signaling mechanism shared with LTD. Nat Neurosci 3(12):1291–1300

Carmichael ST (2012) Brain excitability in stroke: the yin and yang of stroke progression. Arch Neurol 69(2):161–167

Chen ZY, Jing D, Bath KG et al (2006) Genetic variant BDNF (Val66Met) polymorphism alters anxiety-related behavior. Science 314(5796):140–143

Derkach VA, Oh MC, Guire ES, Soderling TR (2007) Regulatory mechanisms of AMPA receptors in synaptic plasticity. Nat Rev Neurosci 8(2):101–113

Dingledine R, Borges K, Bowie D, Traynelis SF (1999) The glutamate receptor ion channels. Pharmacol Rev 51(1):7–61

Duman RS, Aghajanian GK, Sanacora G, Krystal JH (2016) Synaptic plasticity and depression: new insights from stress and rapid-acting antidepressants. Nat Med 22(3):238–249

Ehlers MD (2000) Reinsertion or degradation of AMPA receptors determined by activity-dependent endocytic sorting. Neuron 28(2):511–525

Ellis JS, Zarate CA Jr, Luckenbaugh DA, Furey ML (2014) Antidepressant treatment history as a predictor of response to scopolamine: clinical implications. J Affect Disord 162:39–42
Fleming JJ, England PM (2010) Developing a complete pharmacology for AMPA receptors: a perspective on subtype-selective ligands. Bioorg Med Chem 18(4):1381–1387
Fowler JH, Whalley K, Murray T, O'Neill MJ, McCulloch J (2004) The AMPA receptor potentiator LY404187 increases cerebral glucose utilization and c-fos expression in the rat. J Cereb Blood Flow Metab 24(10):1098–1109
Fukumoto K, Iijima M, Chaki S (2014) Serotonin-1A receptor stimulation mediates effects of a metabotropic glutamate 2/3 receptor antagonist, 2S-2-amino-2-(1S,2S-2-carboxycycloprop-1-yl)-3-(xanth-9-yl)propanoic acid (LY341495), and an N-methyl-D-aspartate receptor antagonist, ketamine, in the novelty-suppressed feeding test. Psychopharmacology 231(11):2291–2298
Gaynes BN, Warden D, Trivedi MH, Wisniewski SR, Fava M, Rush AJ (2009) What did STAR*D teach us? Results from a large-scale, practical, clinical trial for patients with depression. Psychiatr Serv 60(11):1439–1445
Greger IH, Watson JF, Cull-Candy SG (2017) Structural and functional architecture of AMPA-type glutamate receptors and their auxiliary proteins. Neuron 94(4):713–730
Jiang B, Wang F, Yang S et al (2014) SKF83959 produces antidepressant effects in a chronic social defeat stress model of depression through BDNF-TrkB pathway. Int J Neuropsychopharmacol 18(6). https://doi.org/10.1093/ijnp/pyu096
Jourdi H, Hsu YT, Zhou M, Qin Q, Bi X, Baudry M (2009) Positive AMPA receptor modulation rapidly stimulates BDNF release and increases dendritic mRNA translation. J Neurosci 29(27):8688–8697
Kaae BH, Harpsoe K, Kastrup JS et al (2007) Structural proof of a dimeric positive modulator bridging two identical AMPA receptor-binding sites. Chem Biol 14(11):1294–1303
Kato AS, Gill MB, Yu H, Nisenbaum ES, Bredt DS (2010) TARPs differentially decorate AMPA receptors to specify neuropharmacology. Trends Neurosci 33(5):241–248
Kessler RC, Berglund P, Demler O et al (2003) The epidemiology of major depressive disorder: results from the National Comorbidity Survey Replication (NCS-R). JAMA 289(23):3095–3105
Kew JN, Kemp JA (2005) Ionotropic and metabotropic glutamate receptor structure and pharmacology. Psychopharmacology 179(1):4–29
Knapp RJ, Goldenberg R, Shuck C et al (2002) Antidepressant activity of memory-enhancing drugs in the reduction of submissive behavior model. Eur J Pharmacol 440(1):27–35
Koike H, Iijima M, Chaki S (2011) Involvement of AMPA receptor in both the rapid and sustained antidepressant-like effects of ketamine in animal models of depression. Behav Brain Res 224(1):107–111
Kunugi A, Tajima Y, Kuno H, Sogabe S, Kimura H (2018) HBT1, a novel AMPA receptor potentiator with lower agonistic effect, avoided bell-shaped response in in vitro BDNF production. J Pharmacol Exp Ther 364(3):377–389
Kunugi A, Tanaka M, Suzuki A et al (2019) TAK-137, an AMPA-R potentiator with little agonistic effect, has a wide therapeutic window. Neuropsychopharmacology 44(5):961–970
Lepack AE, Fuchikami M, Dwyer JM, Banasr M, Duman RS (2014) BDNF release is required for the behavioral actions of ketamine. Int J Neuropsychopharmacol 18(1). https://doi.org/10.1093/ijnp/pyu033
Li N, Lee B, Liu RJ et al (2010) mTOR-dependent synapse formation underlies the rapid antidepressant effects of NMDA antagonists. Science 329(5994):959–964
Lucassen PJ, Meerlo P, Naylor AS et al (2010) Regulation of adult neurogenesis by stress, sleep disruption, exercise and inflammation: implications for depression and antidepressant action. Eur Neuropsychopharmacol 20(1):1–17
Lynch G (2004) AMPA receptor modulators as cognitive enhancers. Curr Opin Pharmacol 4(1):4–11
Ma Z, Zang T, Birnbaum SG et al (2017) TrkB dependent adult hippocampal progenitor differentiation mediates sustained ketamine antidepressant response. Nat Commun 8(1):1668

Mackowiak M, O'Neill MJ, Hicks CA, Bleakman D, Skolnick P (2002) An AMPA receptor potentiator modulates hippocampal expression of BDNF: an in vivo study. Neuropharmacology 43(1):1–10

Maeng S, Zarate CA Jr, Du J et al (2008) Cellular mechanisms underlying the antidepressant effects of ketamine: role of alpha-amino-3-hydroxy-5-methylisoxazole-4-propionic acid receptors. Biol Psychiatry 63(4):349–352

Majchrzak MJ, Guanowsky V, Freeman J, McLean S, Shaffer CL, Seymour PA (2009) Behavioral pharmacology of the AMPA positive allosteric modulators, PF-4778574 and LY451646. Poster presented at Society for Neuroscience, Pfizer, Chicago, 17-21 October 2009

Malatynska E, Goldenberg R, Shuck L et al (2002) Reduction of submissive behavior in rats: a test for antidepressant drug activity. Pharmacology 64(1):8–17

Marcussen AB, Flagstad P, Kristjansen PE, Johansen FF, Englund U (2008) Increase in neurogenesis and behavioural benefit after chronic fluoxetine treatment in Wistar rats. Acta Neurol Scand 117(2):94–100

Murrough JW, Perez AM, Pillemer S et al (2013) Rapid and longer-term antidepressant effects of repeated ketamine infusions in treatment-resistant major depression. Biol Psychiatry 74(4):250–256

Navarria A, Wohleb ES, Voleti B et al (2015) Rapid antidepressant actions of scopolamine: role of medial prefrontal cortex and M1-subtype muscarinic acetylcholine receptors. Neurobiol Dis 82:254–261

Pinhasov A, Crooke J, Rosenthal D, Brenneman D, Malatynska E (2005) Reduction of submissive behavior model for antidepressant drug activity testing: study using a video-tracking system. Behav Pharmacol 16(8):657–664

Seeburg PH, Hartner J (2003) Regulation of ion channel/neurotransmitter receptor function by RNA editing. Curr Opin Neurobiol 13(3):279–283

Shirayama Y, Hashimoto K (2018) Lack of antidepressant effects of (2R,6R)-hydroxynorketamine in a rat learned helplessness model: comparison with (R)-ketamine. Int J Neuropsychopharmacol 21(1):84–88

Sobolevsky AI, Rosconi MP, Gouaux E (2009) X-ray structure, symmetry and mechanism of an AMPA-subtype glutamate receptor. Nature 462(7274):745–756

Suzuki A, Kunugi A, Tajima Y, Hara H, Yamada R, Kimura H (2018) TAK-653, an AMPA receptor potentiator, produces an antidepressant-like effect with potential fast onset of action like ketamine. Poster presented at 31st The International College of Neuropsychopharmacology CINP 2018, Vienna, 16–19 June 2018; Abstract # PM261: 148–149

Suzuki A et al (2019a) Electrophysiological characterization of a novel AMPA receptor potentiator, TAK-137, in rat hippocampal neurons. Neurosci Lett 712:134488

Suzuki A, Murakami K, Tajima Y et al (2019b) TAK-137, an AMPA receptor potentiator with little agonistic effect, produces antidepressant-like effect without causing psychotomimetic effects in rats. Pharmacol Biochem Behav 183:80–86

Suzuki A et al (2021) Strictly regulated agonist-dependent activation of AMPA-R is the key characteristic of TAK-653 for robust synaptic responses and cognitive improvement. Sci Rep 11:14532

Swiech L, Perycz M, Malik A, Jaworski J (2008) Role of mTOR in physiology and pathology of the nervous system. Biochem Biophys Acta 1784(1):116–132

Tanaka M, Kunugi A, Suzuki A et al (2019) Preclinical characterization of AMPA receptor potentiator TAK-137 as a therapeutic drug for schizophrenia. Pharmacol Res Perspect 7(3):e00479

Voleti B, Navarria A, Liu R-J et al (2013) Scopolamine rapidly increases mammalian target of rapamycin complex 1 signaling, synaptogenesis, and antidepressant behavioral responses. Biol Psychiatry 74(10):742–749

Ward SE, Bax BD, Harries M (2010) Challenges for and current status of research into positive modulators of AMPA receptors. Br J Pharmacol 160(2):181–190

Witkin JM, Overshiner C, Li X et al (2014) M1 and M2 muscarinic receptor subtypes regulate antidepressant-like effects of the rapidly acting antidepressant scopolamine. J Pharmacol Exp Ther 351(2):448–456
Yamada KA (1998) Modulating excitatory synaptic neurotransmission: potential treatment for neurological disease? Neurobiol Dis 5(2):67–80
Yamaguchi JI, Toki H, Qu Y et al (2018) (2R,6R)-Hydroxynorketamine is not essential for the antidepressant actions of (R)-ketamine in mice. Neuropsychopharmacology 43(9):1900–1907
Yu H, Wang DD, Wang Y, Liu T, Lee FS, Chen ZY (2012) Variant brain-derived neurotrophic factor Val66Met polymorphism alters vulnerability to stress and response to antidepressants. J Neurosci 32(12):4092–4101
Yu H, Li M, Zhou D et al (2018) Vesicular glutamate transporter 1 (VGLUT1)-mediated glutamate release and membrane GluA1 activation is involved in the rapid antidepressant-like effects of scopolamine in mice. Neuropharmacology 131:209–222
Zanos P, Gould TD (2018) Mechanisms of ketamine action as an antidepressant. Mol Psychiatry 23(4):801–811
Zarate CA Jr, Singh JB, Carlson PJ et al (2006) A randomized trial of an N-methyl-D-aspartate antagonist in treatment-resistant major depression. Arch Gen Psychiatry 63(8):856–864
Zhou W, Wang N, Yang C, Li XM, Zhou ZQ, Yang JJ (2014) Ketamine-induced antidepressant effects are associated with AMPA receptors-mediated upregulation of mTOR and BDNF in rat hippocampus and prefrontal cortex. Eur Psychiatry 29(7):419–423

AMPA Receptor Potentiators as Potential Rapid-Acting Antidepressants

Emilio Garro-Martínez and Albert Adell

Abstract Major depressive disorder (MDD) is the leading cause of disability worldwide and contributes importantly to the global burden of morbidity according to reports from the World Health Organization. Several types of antidepressant medications are used to treat moderate and severe depressive disorders although weeks or even months are required to produce clinical improvement. However, an estimated one third of the patients have inappropriate responses or no response at all to treatment. Hence, finding new, more efficient and rapid-acting antidepressant therapies is urgently needed. The discovery that ketamine produced rapid and sustained antidepressant effects has been the major breakthrough in the field of pharmacotherapy for MDD. Interestingly, studies with rodents have shown that the antidepressant-like effects of ketamine are dependent on the activation of AMPA receptors (AMPARs) because its behavioral action is blocked by AMPAR antagonists.

Moreover, the removal of specific AMPAR subunits from birth results in behavioral and neurochemical characteristics relevant to depression. Taken together, these findings suggest that the stimulation of AMPARs may be a useful approach to treat MDD. Preclinical studies have reported that positive allosteric modulators (PAMs) – also known as ampakines or AMPAR potentiators – manifest antidepressant-like effects. These molecules bind allosterically to AMPARs and prolong the duration of AMPAR-mediated responses, theoretically without an overstimulation of excitatory transmission. In this chapter, we will review the most important preclinical data to

E. Garro-Martínez
Umeå University, Department of Integrative Medical Biology, Umeå, Sweden

Biomedical Research Networking Center for Mental Health (CIBERSAM), Santander, Spain

A. Adell (✉)
Biomedical Research Networking Center for Mental Health (CIBERSAM), Santander, Spain

Institute of Biomedicine and Biotechnology of de Cantabria, IBBTEC (CSIC-University of Cantabria), Santander, Spain
e-mail: albert.adell@csic.es

K. Hashimoto, M. Manto (eds.), *New Rapid-acting Antidepressants*, Contemporary Clinical Neuroscience,
https://doi.org/10.1007/978-3-030-79790-4_6

date on the antidepressant-like effects of AMPA potentiators discussing their potential use as therapeutic tool in the clinic.

Keywords Glutamate · Positive allosteric modulator · Antidepressant

Introduction

Depression or major depressive disorder (MDD) according to *Diagnostic and Statistical Manual of Mental Disorders, fifth Edition* (DSM-5) is the most common of mental illnesses. It is estimated that around 300 million people suffer from MDD (GBD 2017 Disease and Injury Incidence and Prevalence Collaborators 2018). In 2020, depression reached the ill-fortuned category of the leading cause of disability worldwide and is a major contributor to the overall global burden of disease as stated in the key facts of WHO (30 January 2020). There are interventional therapies of different types that are effective treatments for moderate and severe depression. However, an estimated one third of the patients have inappropriate responses or no response at all to treatment (Fava 2003; Ionescu et al. 2015). Further, even though some response is present, all available antidepressant drugs need to be administered for weeks or months to produce a meaningful clinical improvement (i.e., 50% reduction of severity). Therefore, finding novel rapid-acting antidepressant treatments is a global priority. To meet this goal, some therapies have been developed in recent years. The discovery that ketamine, a *N*-methyl-D-aspartate (NMDA) receptor antagonist, produced rapid (in only 2 h after a single intravenous bolus administration) and sustained (lasting up to 7 days) antidepressant effects (Berman et al. 2000; Zarate Jr et al. 2006; DiazGranados et al. 2010; Singh et al. 2016) transformed completely the field of pharmacotherapy for MDD. Deep brain stimulation (DBS) is another treatment that has shown efficacy in treatment-resistant depression (TRD). Importantly, a number of studies have demonstrated that DBS of the subcallosal cingulate gyrus (Brodmann's area 25, Cg25) or ventral capsule/ventral striatum (VC/VS) is effective in TRD (Mayberg et al. 2005; Malone Jr et al. 2009; Holtzheimer 3rd and Mayberg 2010; Kennedy et al. 2011; Puigdemont et al. 2012). Interestingly, the antidepressant-like effects of both ketamine and DBS need to activate α-amino-3-hydroxy-5-methyl-4-isoxazolepropionic acid (AMPA) receptors (AMPARs) inasmuch as their action is blocked by AMPAR antagonists (Maeng et al. 2008; Li et al. 2010; Koike et al. 2011; Jiménez-Sánchez et al. 2016). Moreover, mice with the AMPAR GluA1 subunit removed – knockout mice – exhibit behavioral and neurochemical characteristics relevant to depression (Kiselycznyk et al. 2015). Taken together, these findings suggest that the stimulation of AMPARs may be a useful approach to treat MDD (Lynch 2006; Alt et al. 2006; O'Neill and Witkin 2007; Arai and Kessler 2007). However, excessive AMPAR stimulation by full AMPAR agonists can lead to psychostimulant and seizure activity (Meldrum 1994). Instead, positive allosteric modulators (PAMs) – also known as ampakines or

AMPAR potentiators – bind allosterically to AMPARs and slow the rate of their deactivation or desensitization (Lynch 2006), thus prolonging the duration of AMPAR-mediated responses theoretically without an overstimulation of excitatory transmission (Lapidus et al. 2013). Although the efficacy of these compounds in preclinical studies of depressive-like behavior has been demonstrated, the transition to clinical trials has not been as fruitful. A look to clinical trials website (www.clinicaltrials.gov) yields one study of CX717 (RespireRx Pharmaceuticals) for attention-deficit/hyperactivity disorder (ADHD, NCT03375021), two studies of safety and tolerability (NCT02561156) and pharmacodynamic activity (NCT03792672) of TAK-653 (Takeda Pharmaceuticals), and one more study examining the add-on therapy of S44475 (Servier) for depression (NCT02805439). So far though, no result has been disclosed.

In this chapter, we will review the most important preclinical data on the antidepressant-like effects of AMPA potentiators with an emphasis in their potential use as therapeutic tool in the clinic.

AMPA Receptors

Glutamate is the principal neurotransmitter in cortical tissue, and its multiple physiological actions can be achieved by binding to ionotropic and metabotropic receptors. Ionotropic receptors can be further subdivided into AMPA, kainate (KA), and NMDA receptors, and they are constructed as assemblies of different subunits, which confers a wide variety of possibilities in terms of interaction with endogenous transmitters, co-transmitters, modulators, and transmembrane proteins. AMPARs are the chief mediators responsible of fast excitatory neurotransmission in the central nervous system (CNS) and collectively with NMDA receptors (NMDARs) can bring about synaptic plasticity. Further to their role in synaptic plasticity, AMPARs are involved in multiple physiological functions such as transfer of information through the synaptic cleft, learning, and memory. On the other hand, alterations of AMPAR function can result in a wide variety of neurological and psychiatric diseases such as epilepsy, pain, depression, or anxiety. Although the important involvement of AMPARs in these physiological and pathological conditions may make them seemingly tantalizing pharmacological targets, their ubiquitous presence throughout the CNS (Beneyto and Meador-Woodruff 2004) suggests that their global stimulation or inhibition by full agonists or antagonists, respectively, could evoke unwanted side effects. In this regard perampanel is the only AMPAR antagonist approved to treat seizures (Zwart et al. 2014). However, it can cause aggressive behavior, irritability, and hostility when overdosed (Brodie et al. 2016).

From a structural viewpoint, AMPARs are composed of four subunits, namely, GluA1–GluA4, which are encoded by the genes *GRIA1–GRIA4* that can be subject to post-transcriptional modifications, such as RNA editing and splicing (Sommer et al. 1990; Rueter et al. 1995). These modifications control the amplitude and duration of excitatory postsynaptic currents (EPSCs). AMPARs can assemble as

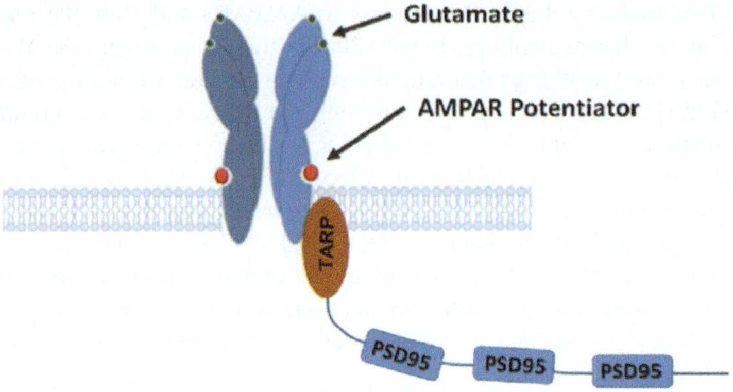

Fig. 1 Schematic representation of an AMPA receptor inserted in the plasma membrane. Each subunit of the AMPA receptor can bind one glutamate molecule and also one molecule of a positive allosteric modulator of the AMPA receptor. The AMPA receptor is anchored to the plasma membrane by auxiliary proteins called transmembrane AMPAR regulatory proteins (TARPs). The C terminus of TARP molecule binds PDZ domain-containing proteins, such as the postsynaptic density protein 95 (PSD95). The interaction between TARP and AMPA receptor subunits appears to be important for the surface delivery of AMPA receptors, whereas the interaction between TARP and PDZ-containing proteins is important for the synaptic targeting of AMPA receptors (Song and Huganir 2002)

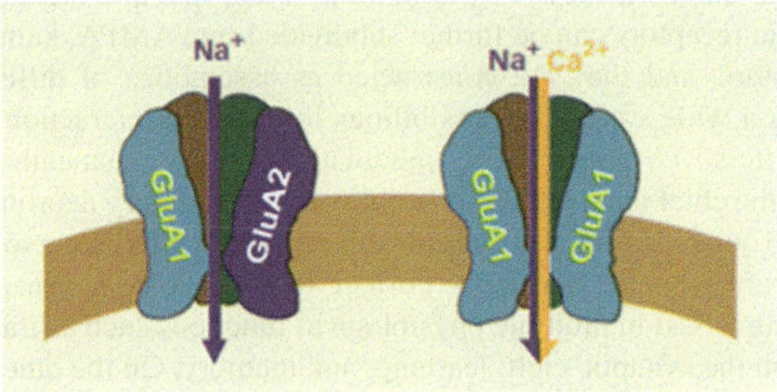

Fig. 2 AMPA receptors containing the GluA2 subunit predominantly allows only the passage of Na^+ ions into the cell. However, AMPA receptors lacking the GluA2 subunit are also permeable to Ca^{2+} ions. (Reproduced from "NIDA. Rare Glutamate Receptor Proliferates After Cocaine Withdrawal". National Institute on Drug Abuse website. https://archives.drugabuse.gov/news-events/nida-notes/2009/12/rare-glutamate-receptor-proliferates-after-cocaine-withdrawal. December 1, 2009. Accessed December 15, 2020)

tetramers of different combinations of the four subunits (Fig. 1) although dimer of dimers that constitute hetero-tetrameric receptors is the most common conformation (Traynelis et al. 2010).

A prominent feature of the subunit composition of AMPARs is that those lacking the GluA2 subunit (mostly containing the GluA1 subunit) are permeable to Ca^{2+} ions (Fig. 2), which makes them good candidates as mediators of cellular and molecular processes leading to synaptic plasticity. In fact, these GluA2-lacking AMPARs provide a mechanism of Ca^{2+} entry into the cell that induces LTP or LTD independent of NMDA receptor activation (Cull-Candy et al. 2006; Liu and Zukin 2007; Lüscher and Malenka 2011). Experimental observations have reported that such robust changes in synaptic transmission can be induced by short tetanic stimulation and can be sustained over hours and even days (Choquet and Hosy 2020).

AMPARs possess a low affinity for glutamate transmitter itself. For this reason, it is of crucial importance that postsynaptic AMPARs are localized closely to glutamate-releasing sites of the presynaptic membrane. In this regard, in addition to subunit composition, other factors contribute to the physiological functioning of AMPARs. For instance, several auxiliary proteins called transmembrane AMPAR regulatory proteins (TARPs) have been identified, which are responsible for anchoring AMPARs to plasma membrane, thus modulating AMPAR trafficking and function. In particular, TARPs determine basic characteristics of AMPAR gating, channel conductance, pharmacology, and expression at synapses (Jackson and Nicoll 2011; Straub and Tomita 2012; Haering et al. 2014). The molecular complex made up of AMPARs and TARPs varies greatly between neuron types and in different regions of the brain (Payne 2008). Overall, the differential expression of AMPAR subunits and the auxiliary proteins results in a considerable variation of AMPAR features and the diversity of the synaptic transfer of information at distinct circuits throughout the brain. For more information about the structural characteristics of AMPARs, the reader is referred to excellent published reviews (Greger et al. 2017; Jacobi and von Engelhardt 2017; Choquet 2018).

AMPA Receptors in Depression

The finding that the NMDA receptor antagonist, ketamine, produced a rapid antidepressant response in MDD patients (Berman et al. 2000; Zarate Jr et al. 2006; DiazGranados et al. 2010; aan het Rot et al. 2010; Singh et al. 2016) turned the attention of the pathophysiology of depression from monoamines to glutamate. In fact, a malfunction of cortico-limbic structures is considered to contribute to MDD (Price and Drevets 2012; Duman et al. 2019; Spellman and Liston 2020), and the communication between different cortical regions and between cortical and subcortical areas is mostly carried out by glutamatergic transmission. Therefore, it was just a matter of time that evidence emerged in favor of the glutamatergic system as a mediator of depressive pathology and a target for the action of rapid-acting antidepressant drugs (Sanacora et al. 2008, 2012; Duman and Aghajanian 2012; Musazzi et al. 2013; Duman et al. 2016; Lener et al. 2017; Murrough et al. 2017). Intriguingly, the rapid antidepressant effects of ketamine depend on the activation of AMPARs by an increased glutamate release in the prefrontal cortex (PFC) (Maeng et al. 2008; Li et al. 2010; Koike et al. 2011), a mechanism shared by another antidepressant therapy such as deep brain stimulation (DBS) (Jiménez-Sánchez et al. 2016).

Postmortem studies revealed altered glutamate levels in brain areas from patients with mood disorders (Hashimoto et al. 2007; Machado-Vieira et al. 2009). In addition, elevated glutamate levels in the occipital cortex and decreased levels in the anterior cingulate cortex appear to be the most consistent findings in nuclear magnetic resonance spectroscopy studies of individuals with mood disorders (Sanacora et al. 2004; Hasler et al. 2007; reviewed in Zarate Jr et al. 2010).

Clinical and preclinical work suggested an implication of AMPAR in MDD although discrepant results were found depending on the brain region and pathological condition. Thus, postmortem studies revealed alterations in the different subunits of the receptor in mood disorders. A downregulation of the GluA1 subunit expression in the striatum was observed in bipolar (but not unipolar) depression (Meador-Woodruff et al. 2001) and in the Brodmann area 24 in MDD (Gibbons et al. 2012). Further, in MDD postmortem tissue, GluA1 and GluA3 mRNAs were diminished in the perirhinal cortex (Beneyto et al. 2007) and in the dentate gyrus and CA1 subareas of the hippocampus (Duric et al. 2013), together with a decrease in the GluA2 and GluA3 mRNAs in the dorsolateral prefrontal cortex although no changes in the protein levels were observed in the latter brain area (Beneyto and Meador-Woodruff 2006). However, no changes in AMPAR binding were observed in other brain areas such as the amygdala or the hippocampus (Noga and Wang 2002; Beneyto et al. 2007).

As it occurs in human postmortem tissue, studies in rodents also exhibited inconsistent results in conditions relevant to depression. For instance, a downregulation of the GluA1 subunit in the hippocampus was found after exposure to acute (Bartanusz et al. 1995) or chronic stress (Toth et al. 2008; Kallarackal et al. 2013), a finding also observed in the PFC (Yuen et al. 2012). However, increases of the GluA1 and GluA2 subunits after acute (Rosa et al. 2002) and increases of only GluA1 after chronic exposure to stress (Schwendt and Jezová 2000; Rosa et al. 2002; Qin et al. 2004) or no change in AMPAR subunit composition (Watanabe et al. 1995; Suenaga et al. 2004; Duric et al. 2013) has been described. These discrepancies could be accounted for by differences in methodological or stress procedures and warrant the need for additional studies leading to a better understanding of the role of AMPARs in mood disorders. Among the different subunits forming the ion channel, the GluA1 subunit appears to be the most relevant to MDD, possibly due to its role in synaptic plasticity (Pittenger and Duman 2008), which is described below.

Effects of Antidepressant Treatments on AMPA Receptors

Classical antidepressants enhance AMPAR function by stimulating both subunit phosphorylation (Svenningsson et al. 2002) and insertion to the synaptic membrane (Martínez-Turrillas et al. 2007; Derkach et al. 2007). Further, stimulation of AMPARs is required for this antidepressant action because their blockade prevents the effects of fluoxetine (Farley et al. 2010), ketamine (Maeng et al. 2008; Li et al.

2010; Koike et al. 2011), lithium (Gould et al. 2008), DBS (Jiménez-Sánchez et al. 2016), and mGluR II antagonists (Karasawa et al. 2005; Koike and Chaki 2014) but not those of imipramine (Li et al. 2010), thus indicating that the effects of tricyclic drugs do not require activation of AMPARs. Further, the systemic and intracortical infusion of (*S*)-AMPA elicits antidepressant-like effects measured as decreased immobility in the forced swim test and increased sucrose consumption (Akinfiresoye and Tizabi 2013; Jiménez-Sánchez et al. 2016).

There is also evidence that chronic treatment with antidepressant drugs promotes the synthesis of the GluA1 subunit (Martínez-Turrillas et al. 2005; Du et al. 2006; Barbon et al. 2011; Réus et al. 2011) although differences may exist with regard to brain regions. For instance, paroxetine and desipramine increased protein levels of GluA1 and GluA2/3 subunits in the hippocampus (Martinez-Turrillas et al. 2002; Martínez-Turrillas et al. 2005), maprotiline raised the GluA1 and GluA2/3 expression in the nucleus accumbens and hippocampus (Tan et al. 2006), fluoxetine increased GluA1 and GluA2 subunits in the retrosplenial granular cortex (Ampuero et al. 2010), and also higher levels of GluA1 were found after acute ketamine administration in the PFC (Li et al. 2010).

History of AMPA Receptor Potentiators

AMPAR potentiators bind to AMPARs in a site, which is different from the glutamate recognition site. When glutamate binds, the AMPAR responds by briefly opening (followed by closing), thus allowing ions to flow into the neuron. If both glutamate and an AMPAR potentiator are bound at the same time, the opening and closing cycle is slightly slowed down, and as a result, more ions flow into the cell and the response of AMPARs is reinforced (Fig. 3). Therefore, the AMPAR

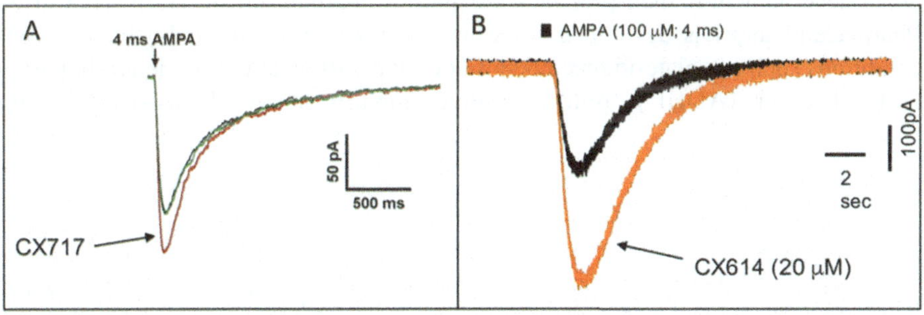

Fig. 3 The initial depolarization produced by AMPA undergoes rapid and almost complete desensitization. (**a**) brief application of AMPA to an individual neuron produces a fast current that is rapidly deactivated. The co-application of the low-impact ampakine, CX717, enhances the amplitude of the AMPA mediated current, but does not affect the deactivation. (**b**) The co-application of the high-impact ampakine, CX614, enhances the amplitude of the AMPA-mediated current but also prolongs the current by slowing the deactivation. (Reproduced from Ampakines® Development Summary – July, 2017, with permission from RespireRx Pharmaceuticals Inc.)

potentiator molecule itself does not directly activate AMPARs. Rather, it interacts in a highly specific manner with the AMPAR, lowering the amount of glutamate required to generate a response and increasing the magnitude and/or duration of the response to any given amount of the transmitter.

Ampakines are a novel class of synthetic compounds that were initially designed to treat the loss of cognition, learning, and memory, biological processes related to aging that were attributed to a weakened glutamatergic transmission (Lynch 2006; Lynch and Gall 2006). Hence, ampakines would facilitate glutamatergic neurotransmission in cortical synapses. In rodents, these compounds also enhance attention and alertness and facilitate learning and memory (Broberg et al. 2009; Hamlyn et al. 2011; Zheng et al. 2011; Baudry et al. 2012). Seminal studies carried out by Gary Lynch at the University of California at Irvine demonstrated that the beneficial effects of ampakines on aging-induced memory impairments persisted long after the drugs were washed out, probably because ampakines selectively enhance the encoding of both short- and long-term memory (Lynch 2002). In general, medications that help to improve intellectual performance are known as *smart drugs*, *cognitive enhancers,* or *nootropics*, even though not all of them potentiate AMPAR function.

From a pharmacological point of view, ampakines are subdivided into two different classes (Arai et al. 2002) known as low-impact (Fig. 4) and high-impact ampakines (Fig. 5). Low-impact ampakines preferentially accelerate channel opening but do not alter receptor desensitization and do not alter agonist binding affinity. However, high-impact ampakines interfere with receptor desensitization and enhance agonist binding. Unlike high-impact ampakines, low-impact ampakines do not seem to produce adverse, pro-convulsant effects even after administration of high doses. Therefore, low-impact ampakines are potentially safe and well tolerated in humans. Indeed, CX717 has entered initial clinical studies in humans (Wesensten et al. 2007; Oertel et al. 2010; Boyle et al. 2012).

At present, only a few drug or biotech companies have AMPAR potentiators under development in their pipelines. RespireRx Pharmaceuticals (formerly Cortex Pharmaceuticals) has developed some ampakines (CX compounds) for different conditions such as opiate-induced respiratory depression, attention-deficit/hyperactivity disorder (ADHD), mild cognitive impairment (MCI) associated with

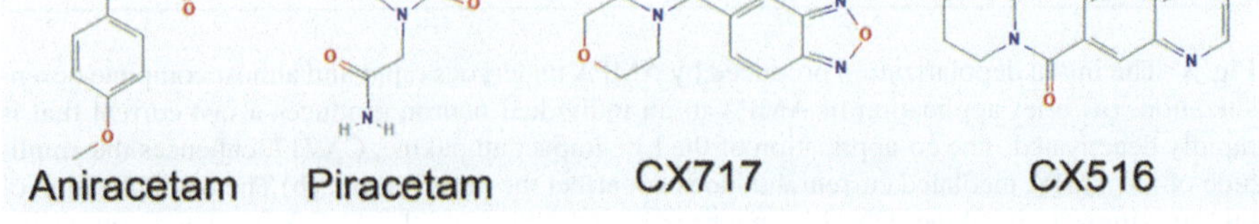

Fig. 4 Several examples of low-impact ampakines

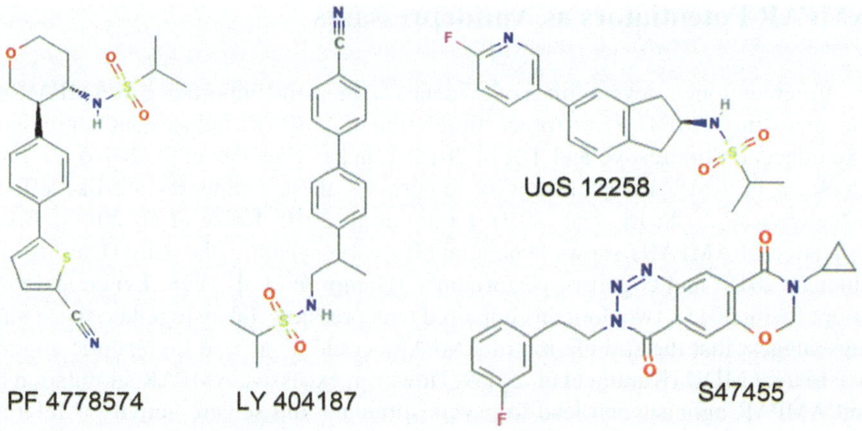

Fig. 5 Several examples of high-impact ampakines

Alzheimer's disease, and fragile X syndrome and also as add-on therapy for schizophrenia. A recent study from our laboratory showed that the low-impact ampakine CX717 evoked antidepressant-like effects in rats (Gordillo-Salas et al. 2020).

Tianeptine was developed by Servier to treat depressive disorder. Somehow counterintuitively, tianeptine was first described as a serotonin uptake potentiator, but a recent paper by Cavalla and co-workers has shown that the drug also facilitates AMPA-mediated glutamatergic transmission (Cavalla et al. 2015). Tulrampator (also known as S47445 or CX1632) is under development by RespireRx Pharmaceuticals and Servier as add-on therapy for the treatment of major depressive disorder (ClinicalTrials.gov Identifier: NCT02805439) and MCI in Alzheimer's disease (ClinicalTrials.gov Identifier: NCT02626572). Tulrampator is a high-impact AMPAR potentiator, which is able to elicit more robust increases in AMPAR activation (Roberts et al. 2010). In 2015 Servier abandoned work on S47445 in depressive patients with the degenerative disease on efficacy grounds. Recent findings have also evidenced antidepressant-like effects of S47445 in rodents (Mendez-David et al. 2017; Pilar-Cuellar et al. 2019). Takeda also has an AMPAR potentiator, TAK-653, which is examined as treatment for depressive disorders (ClinicalTrials.gov Identifier: NCT03792672). Takeda had previously discontinued work on a previous molecule, TAK-137, for safety reasons. In animals, high-impact AMPAR potentiators enhance cognition and memory at low doses, but produce motor coordination disruptions, convulsions, and neurotoxicity at higher doses (Ranganathan et al. 2017). This suggests that their therapeutic window is very narrow and, hence, greater discrimination in activity is needed before any clinical application.

AMPAR Potentiators as Antidepressants

As aforementioned, several lines of evidence suggest the importance of AMPARs in the treatment of MDD. *First*, direct stimulation of AMPARs induces antidepressant-like effects (Akinfiresoye and Tizabi 2013; Jiménez-Sánchez et al. 2016). *Second*, blockade of AMPARs attenuates or suppresses these antidepressant-like effects (Maeng et al. 2008; Li et al. 2010; Farley et al. 2010; Koike et al. 2011). *Third*, activation of AMPARs shows beneficial effects on synaptic plasticity (Diering and Huganir 2018) and cognitive performance (Hampson et al. 1998; Lynch and Gall 2006; Partin 2015), two domains impaired in depression. Taken together, these findings suggest that the stimulation of AMPARs could be a good therapeutic alternative to treat MDD (Kunugi et al. 2019). However, excessive AMPAR stimulation by full AMPAR agonists can lead to psychostimulant and seizure activity (Meldrum 1994). Instead, positive allosteric modulators (PAMs) – also known as ampakines – bind allosterically to AMPARs and slow the rate of their deactivation or desensitization (Lynch 2006), thus prolonging the duration of AMPAR-mediated responses (Lapidus et al. 2013). Because ampakines do not activate directly AMPARs, the drugs are only effective with an underlying glutamatergic transmission (Arai et al. 1996). For this reason, it was hypothesized that ampakines would have therapeutic potential as antidepressant drugs (Alt et al. 2005; Alt et al. 2006; Bleakman et al. 2007; O'Neill and Witkin 2007).

One of the important hints in favor of the potential antidepressant action of AMPAR potentiators is the fact that these compounds have the ability to increase the synthesis of brain-derived neurotrophic factor (BDNF) (Zafra et al. 1990; Lindefors et al. 1992; Hayashi et al. 1999; Lauterborn et al. 2000; Legutko et al. 2001; Mackowiak et al. 2002; Radin et al. 2018; Shen et al. 2019; Gordillo-Salas et al. 2020), a trophic factor that is needed for the manifestation of antidepressant effects after effective treatments (Vaidya and Duman 2001; Castrén and Rantamäki 2010a, b). Furthermore, some types of stress, a common precipitating factor for depression habitually used in animal experiments, caused atrophy and even cell death in the hippocampus (Magariños et al. 1996; Sheline et al., 1996; McEwen 2004), alterations that correlate with a decrease in BDNF in the same brain region (Duman et al. 1997; Duman and Monteggia 2006) and are reverted by antidepressant treatment (Duman 2004; Duman et al. 1997). The role of BDNF in depression and its treatment received further support from human postmortem studies that revealed that brain tissue from depressed subjects chronically treated with antidepressants had increased BDNF levels (Chen et al. 2001a) in comparison with the reduced BDNF level postmortem brains and plasma from depressed patients (Dwivedi et al. 2003; Kim et al. 2007).

Another cellular mechanism that is associated with antidepressant effects is the proliferation of progenitor cells in the dentate gyrus of the hippocampus, a process called neurogenesis, which is engaged by different antidepressant strategies

(Encinas et al. 2006; Pascual-Brazo et al. 2012; Ferrés-Coy et al. 2013) and depends also on BDNF expression. This indicates the necessity of hippocampal neurogenesis for the behavioral effects of chronic antidepressants (Santarelli et al. 2003). It has been shown that only 24 hours after the administration of an AMPAR potentiator, there is a significant upregulation of hippocampal proliferation markers (Bai et al. 2003), a procedure not observed after conventional antidepressant compounds, which suggests that AMPAR potentiators could exhibit a rapid onset of antidepressant action (Knapp et al. 2002).

Early preclinical studies showed that the acute administration of LY392098 produced antidepressant-like responses in the forced swim and the tail suspension tests (Li et al. 2001), two of the most used behavioral paradigms relevant to depression with a very high predictive validity. These effects were AMPAR-dependent because they were blocked by pretreating the animals with GYKI53655, a noncompetitive antagonist of the AMPAR. Similar results in the FST and the sucrose preference test were obtained after chronic administration of LY392098 after exposure of the animals to chronic unpredictable mild stress (CUMS) (Farley et al. 2010). These effects were corroborated by other AMPAR potentiators such as LY451646 (Bai et al. 2003; Li et al. 2003; Andreasen et al. 2015), piracetam and aniracetam (Knapp et al. 2002), and CX717 (Gordillo-Salas et al. 2020). Other recent works showed antidepressant-like effects of S47445 in olfactory bulbectomized mice (Pilar-Cuellar et al. 2019) and PF-4778574 in CUMS-exposed mice (Shen et al. 2019). S47455 also exhibited anxiolytic and antidepressant properties in the chronic corticosterone and the CUMS models relevant to depression, behavioral effects that were associated with a parallel increase of BDNF expression in the hippocampus (Mendez-David et al. 2017). Behaviorally effective doses of LY392098 produced no increase in dialysate levels of noradrenaline, dopamine, or serotonin (Skolnick et al. 2001). However, the acute administration of CX717 produced increases in the release of noradrenaline, dopamine, and serotonin, but only when this compound was infused directly in the PFC (Gordillo-Salas et al. 2020). The rapid antidepressant-like effects of the systemic administration of these compounds did not appear to be associated with increased output of monoamines. It is possible that these AMPAR potentiators penetrate poorly into the brain, not sufficiently to increase the release of these transmitters. Interestingly, unlike the perfusion of a full agonist, which increases the output of serotonin, dopamine, and glutamate in the PFC (Jedema and Moghaddam 1994; Jiménez-Sánchez et al. 2016), CX717 increases monoamine, but not glutamate, output (Gordillo-Salas et al. 2020). The absence of increased glutamate release could be responsible for the lack of seizure activity after administration of CX717 (Ren et al. 2012) and suggests that AMPAR potentiators would exert a more efficient utilization of pre-existing synaptic glutamate, by keeping the AMPAR channel open. It remains to be determined whether these effects are shared by other AMPAR potentiators or are particular of CX717.

Intracellular Signaling of AMPA Receptor Potentiators

There is compelling evidence of a relationship between cAMP-response-element-binding protein (CREB) and depressive behaviors. However, activation (phosphorylation) of CREB can either attenuate or evoke depressive-like behaviors in rodents depending on the brain region. Hence, phosphorylation of CREB in the hippocampus mediates antidepressant effects, whereas the same action in the nucleus accumbens (NAc) produces various depressive-like effects in rodents (Barrot et al. 2002; Pliakas et al. 2001; Chen et al. 2001b). In support of these premises, it has been observed that a wide variety of conventional antidepressant treatments (e.g., noradrenaline reuptake inhibitors, selective serotonin reuptake inhibitors, and electroconvulsive therapy) increase CREB activity within the hippocampus (Nibuya et al. 1996; Jeon et al. 1997; Thome et al. 2000).

As aforementioned, further work has also implicated neurotrophic factors such as BDNF in the antidepressant effects of standard antidepressant drugs (Autry et al. 2011; Björkholm and Monteggia 2016) and the AMPAR potentiator PF-4778574 (Shen et al. 2019). In addition, the neural plasticity associated with the rapid antidepressant effects (synaptogenesis), which happens to be reduced in the pathological state, is dependent on AMPAR, BDNF, and mechanistic target of rapamycin (mTOR) signaling (Li et al. 2010; Liu et al. 2012; Lepack et al. 2014; Duman et al. 2016). Several investigations have further demonstrated that the engagement of AMPARs is essential for the activation of the mTOR pathway (Hashimoto 2009; Li et al. 2010; Zhou et al. 2014; Zanos et al. 2016), thus suggesting a role of AMPAR potentiators as compounds with potential antidepressant properties. Overall, this reinforces the well-known role of these proteins in the etiopathological mechanisms of depression as well as in their therapeutic effects (Duman et al. 1997; Krishnan and Nestler 2008).

In vitro and ex vivo studies have shown that the activation of AMPARs upregulates the expression of BDNF in the hippocampus (Zafra et al. 1990; Lindefors et al. 1992) and BDNF expression is regulated by CREB (Duman 1998; Carlezon Jr et al. 2005). In line with this previous work, AMPA potentiators exhibit the same ability (Lauterborn et al. 2000; Skolnick et al. 2001; Legutko et al. 2001; Mackowiak et al. 2002; Radin et al. 2018; Shen et al. 2019; Gordillo-Salas et al. 2020). As depicted in Fig. 6, two different mechanisms have been postulated to be triggered by the activation of AMPARs by AMPAR potentiators: (*i*) intracellular entry of Ca^{2+} ions through L-type voltage-dependent Ca^{2+} channels (VDCC) (Zafra et al. 1990; Lauterborn et al. 2000) and (*ii*) Ca^{2+}-independent mitogen-activated protein kinase (MAPK) pathway (Hayashi et al. 1999). Both pathways lead to the activation of CREB (Xing et al. 1996) and subsequent increment of BDNF levels (Xing et al. 1996; Blendy 2006). Finally, the AMPAR potentiator-induced increase in intracellular Ca^{2+} can also induce the release of BDNF (Alt et al. 2006; Jourdi et al. 2009; Duman and Voleti 2012). The binding of BDNF to its specific receptor tropomyosin receptor kinase B (TrkB) can prompt the intracellular synthesis of ERK or Akt, both procedures leading to the activation of mTOR. The downstream endpoint of all

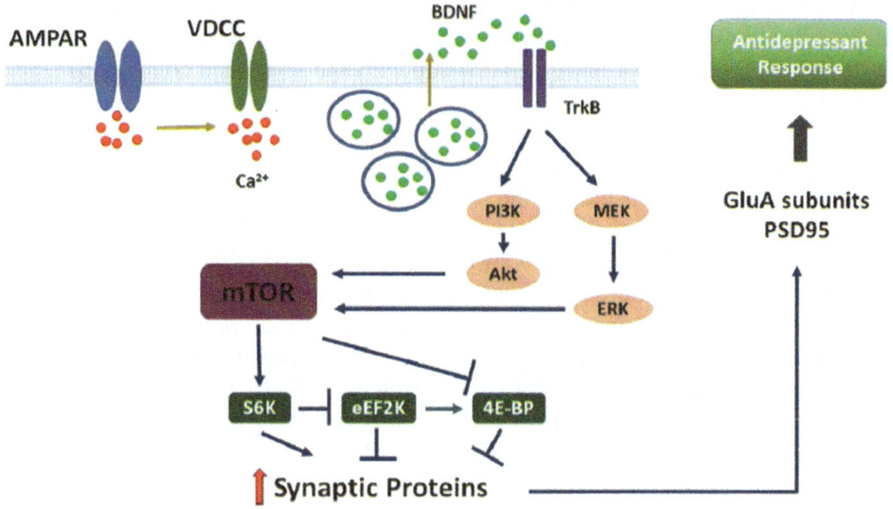

Fig. 6 Signaling pathways engaged by the activation of AMPARs. The entry of Ca^{2+} after stimulation of GluA2-lacking AMPARs induces a rapid depolarization, which appears to lead to the activation of voltage-dependent Ca^{2+} channels (VDCC) and activity-dependent release of brain-derived neurotrophic factor (BDNF). BDNF then binds to tropomyosin-related kinase B (TrkB) receptors, which produces the activation of downstream signaling pathways [phosphatidyl inositol-3 kinase (PI3K)–Akt and Ras–mitogen-activated protein kinase (MAPK)]. These pathways stimulate mechanistic/mammalian target of rapamycin (mTOR) signaling, which elevates S6 kinase (S6K, a member of the ribosomal S6 kinase family) and the synthesis of synaptic proteins such as postsynaptic density-95 (PSD95) and the AMPA receptor GluA1 subunit. GluA1-containing AMPA receptors are eventually inserted into the plasma membrane, which contributes to synaptogenesis and antidepressant action. Abbreviations: 4E-BP, 4E binding protein; eEF2K, eukaryotic elongation factor-2 kinase; ERK, extracellular signal-regulated kinase; MEK, MAP/ERK kinase; PSD95, postsynaptic density protein 95

these pathways is the translocation of CREB into the nucleus and the synthesis of synaptic proteins such as PSD95. Unlike ketamine, the administration of the ampakine CX717 does not seem to eventually increase the synthesis of the GluA1 subunit in naïve rats. Nevertheless, PF-4778574 increased membrane GluA1 in CUMS-exposed mice in a dose-dependent manner (Shen et al. 2019). It appears that the activation of AMPAR transmission by ketamine is carried out by an increased glutamate release (López-Gil et al. 2019), whereas in the case of CX717, the allosteric activation of AMPARs is produced in absence of glutamate release (Gordillo-Salas et al. 2020). It remains to be determined whether this is an effect particular of CX717 or common to other ampakines. Possibly related to the lack of effect on glutamate release, CX717 was also unable to potentiate the synthesis of GluA1 (Gordillo-Salas et al. 2020), which is characteristic of other rapid-acting antidepressants such as ketamine (Li et al. 2010). It is possible that AMPAR potentiators are able to stimulate AMPAR signaling without modifying the structure, subunit composition, or membrane insertion of AMPARs. Alternatively, AMPAR potentiators may be able to increase the synthesis of GluA1, but only when its levels are substantially reduced as a consequence of stress exposure (Shen et al. 2019).

Another protein with a strong relationship with antidepressant effects is p11, also known as S100 calcium-binding protein A10 (S100A10). p11 plays an important role in depression as well as in the responses to antidepressant treatment (Svenningsson et al. 2013). Also, mice with p11 deletion show a depressive-like behavior, an effect abolished by the overexpression of p11 in the nucleus accumbens (Warner-Schmidt et al. 2012). p11 is upregulated by antidepressant drugs (Svenningsson et al. 2006) most likely through the binding of BDNF to TrkB receptor (Warner-Schmidt et al. 2010). On the other hand, p11 has a bidirectional relationship with serotonin transmission in that p11 potentiates serotonin neurotransmission (Svenningsson et al. 2013) and serotonin controls p11 expression (Svenningsson et al. 2006; Warner-Schmidt et al. 2010). Unfortunately, there is no information on the effects of AMPAR potentiators on the levels of p11. Only one work has reported a sustained synthesis of p11 after a single dose of CX717 (Gordillo-Salas et al. 2020).

In summary, the increased synthesis of BDNF would activate mTOR and CREB signaling downstream eventually evoking the formation of new synaptic proteins, a process prevented by an AMPAR receptor blocker (Jourdi et al. 2009). The elaboration of new synapses in the prefrontal cortex is likely important for the antidepressant-like effects of AMPAR potentiators. Furthermore, the increase of PSD95 might be indicative of the formation of new synaptic contacts at this time point (Nikonenko et al. 2008). Further research is needed to discern the exact mechanism of the participation of p11.

Conclusion

Glutamate is the most prevalent excitatory transmitter in the brain, and the glutamate connections in the PFC are involved in many higher brain functions, such as perception, attention, memory, language, consciousness, affect, etc. It also plays a key role in cognitive processes, such as working (short-term) memory and executive functions as well as in action planning, decision-taking, and the processing of emotional signals. AMPAR potentiators could have positive effects on these functions by selectively boosting glutamate activity. It appears that such drugs can also exert a rapid-acting antidepressant action, but the duration of the effect may depend on the particular compound. Although the precise mechanism of action is not fully understood, it seems that the rapid antidepressant effects cannot be accounted for by increased extracellular levels of glutamate and/or monoamines, but rather due to a rapid compensatory synthesis of BDNF and mTOR. BDNF is a major regulator of synaptic plasticity that is controlled upstream by AMPARs together with NMDA receptors. An important role for p11 has been also hypothesized, but the limited data available needs further replication with different compounds. Although there are no human trials on AMPAR potentiators/ampakines and depression currently, it seems to be a promising avenue for future research based upon the positive data coming from preclinical studies.

Acknowledgments This work was supported by the Instituto de Salud Carlos III, Subdirección General de Evaluación y Fomento de la Investigación (FIS Grant PI19/00170) that was co-funded by the European Regional Development Fund ("A way to build Europe"). Funding from the Centro de Investigación Biomédica en Red de Salud Mental (CIBERSAM), Instituto de Salud Carlos III is also acknowledged. The funding agencies had no role in the design and conduct of the study, collection, management, analyses, and interpretation of the data and preparation, review, or approval of the manuscript and the decision to submit it for publication. *Conflict of interest:* The authors have no conflict of interest.

References

aan het Rot M, Collins KA, Murrough JW, Perez AM, Reich DL, Charney DS, Mathew SJ (2010) Safety and efficacy of repeated-dose intravenous ketamine for treatment-resistant depression. Biol Psychiatry 67:139–145. https://doi.org/10.1016/j.biopsych.2009.08.038

Akinfiresoye L, Tizabi Y (2013) Antidepressant effects of AMPA and ketamine combination: role of hippocampal BDNF, synapsin, and mTOR. Psychopharmacology 230:291–298. https://doi.org/10.1007/s00213-013-3153-2

Alt A, Nisenbaum ES, Bleakman D, Witkin JM (2006) A role for AMPA receptors in mood disorders. Biochem Pharmacol 71:1273–1288. https://doi.org/10.1016/j.bcp.2005.12.022

Alt A, Witkin JM, Bleakman D (2005) AMPA receptor potentiators as novel antidepressants. Curr Pharm Des 11:1511–1527. https://doi.org/10.2174/1381612053764814

Ampuero E, Rubio FJ, Falcon R, Sandoval M, Diaz-Veliz G, Gonzalez RE, Earle N, Dagnino-Subiabre A, Aboitiz F, Orrego F, Wyneken U (2010) Chronic fluoxetine treatment induces structural plasticity and selective changes in glutamate receptor subunits in the rat cerebral cortex. Neuroscience 169:98–108. https://doi.org/10.1016/j.neuroscience.2010.04.035

Andreasen JT, Fitzpatrick CM, Larsen M, Skovgaard L, Nielsen SD, Clausen RP, Troelsen K, Pickering DS (2015) Differential role of AMPA receptors in mouse tests of antidepressant and anxiolytic action. Brain Res 1601:117–126. https://doi.org/10.1016/j.brainres.2015.01.001

Arai AC, Xia YF, Rogers G, Lynch G, Kessler M (2002) Benzamide-type AMPA receptor modulators form two subfamilies with distinct modes of action. J Pharmacol Exp Ther 303:1075–1085. https://doi.org/10.1124/jpet.102.040360

Arai AC, Kessler M (2007) Pharmacology of ampakine modulators: from AMPA receptors to synapses and behavior. Curr Drug Targets 8:583–602. https://doi.org/10.2174/138945007780618490

Arai A, Guidotti A, Costa E, Lynch G (1996) Effect of the AMPA receptor modulator IDRA 21 on LTP in hippocampal slices. Neuroreport 7:2211–2215. https://doi.org/10.1097/00001756-199609020-00031

Autry AE, Adachi M, Nosyreva E, Na ES, Los MF, Cheng PF, Kavalali ET, Monteggia LM (2011) NMDA receptor blockade at rest triggers rapid behavioural antidepressant responses. Nature 475:91–95. https://doi.org/10.1038/nature10130

Bai F, Bergeron M, Nelson DL (2003) Chronic AMPA receptor potentiator (LY451646) treatment increases cell proliferation in adult rat hippocampus. Neuropharmacology 44:1013–1021. https://doi.org/10.1016/s0028-3908(03)00104-7

Barbon A, Caracciolo L, Orlandi C, Musazzi L, Mallei A, La Via L, Bonini D, Mora C, Tardito D, Gennarelli M, Racagni G, Popoli M, Barlati S (2011) Chronic antidepressant treatments induce a time-dependent up-regulation of AMPA receptor subunit protein levels. Neurochem Int 59:896–905. https://doi.org/10.1016/j.neuint.2011.07.013

Barrot M, Olivier JD, Perrotti LI, DiLeone RJ, Berton O, Eisch AJ, Impey S, Storm DR, Neve RL, Yin JC, Zachariou V, Nestler EJ (2002) CREB activity in the nucleus accumbens shell controls gating of behavioral responses to emotional stimuli. Proc Natl Acad Sci U S A 99:11435–11440. https://doi.org/10.1073/pnas.172091899

Bartanusz V, Aubry JM, Pagliusi S, Jezova D, Baffi J, Kiss JZ (1995) Stress-induced changes in messenger RNA levels of N-methyl-D-aspartate and AMPA receptor subunits in selected regions of the rat hippocampus and hypothalamus. Neuroscience 66:247–252. https://doi.org/10.1016/0306-4522(95)00084-v

Baudry M, Kramar E, Xu X, Zadran H, Moreno S, Lynch G, Gall C, Bi X (2012) Ampakines promote spine actin polymerization, long-term potentiation, and learning in a mouse model of Angelman syndrome. Neurobiol Dis 47:210–215. https://doi.org/10.1016/j.nbd.2012.04.002

Beneyto M, Meador-Woodruff JH (2004) Expression of transcripts encoding AMPA receptor subunits and associated postsynaptic proteins in the macaque brain. J Comp Neurol 468:530–554. https://doi.org/10.1002/cne.10981

Beneyto M, Kristiansen LV, Oni-Orisan A, McCullumsmith RE, Meador-Woodruff JH (2007) Abnormal glutamate receptor expression in the medial temporal lobe in schizophrenia and mood disorders. Neuropsychopharmacology 32:1888–1902. https://doi.org/10.1038/sj.npp.1301312

Beneyto M, Meador-Woodruff JH (2006) Lamina-specific abnormalities of AMPA receptor trafficking and signaling molecule transcripts in the prefrontal cortex in schizophrenia. Synapse 60:585–598. https://doi.org/10.1002/syn.20329

Berman RM, Cappiello A, Anand A, Oren DA, Heninger GR, Charney DS, Krystal JH (2000) Antidepressant effects of ketamine in depressed patients. Biol Psychiatry 47:351–354. https://doi.org/10.1016/s0006-3223(99)00230-9

Björkholm C, Monteggia LM (2016) BDNF - a key transducer of antidepressant effects. Neuropharmacology 102:72–79. https://doi.org/10.1016/j.neuropharm.2015.10.034

Bleakman D, Alt A, Witkin JM (2007) AMPA receptors in the therapeutic management of depression. CNS Neurol Disord Drug Targets 6:117–126. https://doi.org/10.2174/187152707780363258

Blendy JA (2006) The role of CREB in depression and antidepressant treatment. Biol Psychiatry 59:1144–1150. https://doi.org/10.1016/j.biopsych.2005.11.003

Boyle J, Stanley N, James LM, Wright N, Johnsen S, Arbon EL, Dijk DJ (2012) Acute sleep deprivation: the effects of the AMPAKINE compound CX717 on human cognitive performance, alertness and recovery sleep. J Psychopharmacol 26:1047–1057. https://doi.org/10.1177/0269881111405353

Broberg BV, Glenthøj BY, Dias R, Larsen DB, Olsen CK (2009) Reversal of cognitive deficits by an ampakine (CX516) and sertindole in two animal models of schizophrenia--sub-chronic and early postnatal PCP treatment in attentional set-shifting. Psychopharmacology 206:631–640. https://doi.org/10.1007/s00213-009-1540-5

Brodie MJ, Besag F, Ettinger AB, Mula M, Gobbi G, Comai S, Aldenkamp AP, Steinhoff BJ (2016) Epilepsy, antiepileptic drugs, and aggression: an evidence-based review. Pharmacol Rev 68:563–602. https://doi.org/10.1124/pr.115.012021

Carlezon WA Jr, Duman RS, Nestler EJ (2005) The many faces of CREB. Trends Neurosci 28:436–445. https://doi.org/10.1016/j.tins.2005.06.005

Castrén E, Rantamäki T (2010a) The role of BDNF and its receptors in depression and antidepressant drug action: reactivation of developmental plasticity. Dev Neurobiol 70:289–297. https://doi.org/10.1002/dneu.20758

Castrén E, Rantamäki T (2010b) Role of brain-derived neurotrophic factor in the aetiology of depression: implications for pharmacological treatment. CNS Drugs 24:1–7. https://doi.org/10.2165/11530010-000000000-00000

Cavalla D, Chianelli F, Korsak A, Hosford PS, Gourine AV, Marina N (2015) Tianeptine prevents respiratory depression without affecting analgesic effect of opiates in conscious rats. Eur J Pharmacol 761:268–272. https://doi.org/10.1016/j.ejphar.2015.05.067

Chen B, Dowlatshahi D, MacQueen GM, Wang JF, Young LT (2001a) Increased hippocampal BDNF immunoreactivity in subjects treated with antidepressant medication. Biol Psychiatry 50:260–265. https://doi.org/10.1016/s0006-3223(01)01083-6

Chen AC, Shirayama Y, Shin KH, Neve RL, Duman RS (2001b) Expression of the cAMP response element binding protein (CREB) in hippocampus produces an antidepressant effect. Biol Psychiatry 49:753–762. https://doi.org/10.1016/s0006-3223(00)01114-8

Choquet D (2018) Linking nanoscale dynamics of AMPA receptor organization to plasticity of excitatory synapses and learning. J Neurosci 38:9318–9329. https://doi.org/10.1523/JNEUROSCI.2119-18.2018

Choquet D, Hosy E (2020) AMPA receptor nanoscale dynamic organization and synaptic plasticities. Curr Opin Neurobiol 63:137–145. https://doi.org/10.1016/j.conb.2020.04.003

Cull-Candy S, Kelly L, Farrant M (2006) Regulation of Ca^{2+}-permeable AMPA receptors: synaptic plasticity and beyond. Curr Opin Neurobiol 16:288–297. https://doi.org/10.1016/j.conb.2006.05.012

Derkach VA, Oh MC, Guire ES, Soderling TR (2007) Regulatory mechanisms of AMPA receptors in synaptic plasticity. Nat Rev Neurosci 8:101–113. https://doi.org/10.1038/nrn2055

DiazGranados N, Ibrahim LA, Brutsche NE, Ameli R, Henter ID, Luckenbaugh DA, Machado-Vieira R, Zarate CA Jr (2010) Rapid resolution of suicidal ideation after a single infusion of a *N*-methyl-D-aspartate antagonist in patients with treatment-resistant major depressive disorder. J Clin Psychiatry 71:1605–1611. https://doi.org/10.4088/JCP.09m05327blu

Diering GH, Huganir RL (2018) The AMPA receptor code of synaptic plasticity. Neuron 100:314–329. https://doi.org/10.1016/j.neuron.2018.10.018

Du J, Machado-Vieira R, Maeng S, Martinowich K, Manji HK, Zarate CA Jr (2006) Enhancing AMPA to NMDA throughput as a convergent mechanism for antidepressant action. Drug Discov Today Ther Strateg 3:519–526. https://doi.org/10.1016/j.ddstr.2006.11.012

Duman RS, Aghajanian GK (2012) Synaptic dysfunction in depression: potential therapeutic targets. Science 338:68–72. https://doi.org/10.1126/science.1222939

Duman RS, Aghajanian GK, Sanacora G, Krystal JH (2016) Synaptic plasticity and depression: new insights from stress and rapid-acting antidepressants. Nat Med 22(3):238–249. https://doi.org/10.1038/nm.4050

Duman RS, Sanacora G, Krystal JH (2019) Altered connectivity in depression: GABA and glutamate neurotransmitter deficits and reversal by novel treatments. Neuron 102:75–90. https://doi.org/10.1016/j.neuron.2019.03.013

Duman RS, Heninger GR, Nestler EJ (1997) A molecular and cellular theory of depression. Arch Gen Psychiatry 54:597–606. https://doi.org/10.1001/archpsyc.1997.01830190015002

Duman RS, Monteggia LM (2006) A neurotrophic model for stress-related mood disorders. Biol Psychiatry 59:1116–1127. https://doi.org/10.1016/j.biopsych.2006.02.013

Duman RS (2004) Role of neurotrophic factors in the etiology and treatment of mood disorders. NeuroMolecular Med 5:11–25. https://doi.org/10.1385/NMM:5:1:011

Duman RS (1998) Novel therapeutic approaches beyond the serotonin receptor. Biol Psychiatry 44:324–335. https://doi.org/10.1016/s0006-3223(98)00031-6

Duman RS, Voleti B (2012) Signaling pathways underlying the pathophysiology and treatment of depression: novel mechanisms for rapid-acting agents. Trends Neurosci 35:47–56. https://doi.org/10.1016/j.tins.2011.11.004

Duric V, Banasr M, Stockmeier CA, Simen AA, Newton SS, Overholser JC, Jurjus GJ, Dieter L, Duman RS (2013) Altered expression of synapse and glutamate related genes in post-mortem hippocampus of depressed subjects. Int J Neuropsychopharmacol 16:69–82. https://doi.org/10.1017/S1461145712000016

Dwivedi Y, Rizavi HS, Conley RR, Roberts RC, Tamminga CA, Pandey GN (2003) Altered gene expression of brain-derived neurotrophic factor and receptor tyrosine kinase B in postmortem brain of suicide subjects. Arch Gen Psychiatry 60:804–815. https://doi.org/10.1001/archpsyc.60.8.804

Encinas JM, Vaahtokari A, Enikolopov G (2006) Fluoxetine targets early progenitor cells in the adult brain. Proc Natl Acad Sci U S A 103:8233–8238. https://doi.org/10.1073/pnas.0601992103
Farley S, Apazoglou K, Witkin JM, Giros B, Tzavara ET (2010) Antidepressant-like effects of an AMPA receptor potentiator under a chronic mild stress paradigm. Int J Neuropsychopharmacol 13:1207–1218. https://doi.org/10.1017/S1461145709991076
Fava M (2003) Diagnosis and definition of treatment-resistant depression. Biol Psychiatry 53:649–659. https://doi.org/10.1016/s0006-3223(03)00231-2
Ferrés-Coy A, Pilar-Cuellar F, Vidal R, Paz V, Masana M, Cortés R, Carmona MC, Campa L, Pazos A, Montefeltro A, Valdizán EM, Artigas F, Bortolozzi A (2013) RNAi-mediated serotonin transporter suppression rapidly increases serotonergic neurotransmission and hippocampal neurogenesis. Transl Psychiatry 3:e211. https://doi.org/10.1038/tp.2012.135
GBD 2017 Disease and Injury Incidence and Prevalence Collaborators (2018) Global, regional, and national incidence, prevalence, and years lived with disability for 354 diseases and injuries for 195 countries and territories, 1990–2017: a systematic analysis for the Global Burden of Disease Study 2017. Lancet 392:1789–1858. https://doi.org/10.1016/S0140-6736(18)32279-7
Gibbons AS, Brooks L, Scarr E, Dean B (2012) AMPA receptor expression is increased postmortem samples of the anterior cingulate from subjects with major depressive disorder. J Affect Disord 136:1232–1237. https://doi.org/10.1016/j.jad.2011.10.001
Gordillo-Salas M, Pascual-Antón R, Ren J, Greer J, Adell A (2020) Antidepressant-like effects of CX717, a positive allosteric modulator of AMPA receptors. Mol Neurobiol 57:3498–3507. https://doi.org/10.1007/s12035-020-01954-x
Gould TD, O'Donnell KC, Dow ER, Du J, Chen G, Manji HK (2008) Involvement of AMPA receptors in the antidepressant-like effects of lithium in the mouse tail suspension test and forced swim test. Neuropharmacology 54:577–587. https://doi.org/10.1016/j.neuropharm.2007.11.002
Greger IH, Watson JF, Cull-Candy SG (2017) Structural and functional architecture of AMPA-type glutamate receptors and their auxiliary proteins. Neuron 94:713–730. https://doi.org/10.1016/j.neuron.2017.04.009
Haering SC, Tapken D, Pahl S, Hollmann M (2014) Auxiliary subunits: shepherding AMPA receptors to the plasma membrane. Membranes (Basel) 4:469–490. https://doi.org/10.3390/membranes4030469
Hamlyn E, Brand L, Shahid M, Harvey BH (2011) The ampakine, Org 26576, bolsters early spatial reference learning and retrieval in the Morris water maze: a subchronic, dose-ranging study in rats. Behav Pharmacol 20:662–667. https://doi.org/10.1097/FBP.0b013e328331ba1b
Hampson RE, Rogers G, Lynch G, Deadwyler SA (1998) Facilitative effects of the ampakine CX516 on short-term memory in rats: enhancement of delayed-nonmatch-to-sample performance. J Neurosci 18:2740–2747. https://doi.org/10.1523/JNEUROSCI.18-07-02740.1998
Hashimoto K, Sawa A, Iyo M (2007) Increased levels of glutamate in brains from patients with mood disorders. Biol Psychiatry 62:1310–1316. https://doi.org/10.1016/j.biopsych.2007.03.017
Hashimoto K (2009) Emerging role of glutamate in the pathophysiology of major depressive disorder. Brain Res Rev 61:105–123. https://doi.org/10.1016/j.brainresrev.2009.05.005
Hasler G, Van Der Veen JW, Tumonis T, Meyers N, Shen J, Drevets WC (2007) Reduced prefrontal glutamate/glutamine and gamma-aminobutyric acid levels in major depression determined using proton magnetic resonance spectroscopy. Arch Gen Psychiatry 64:193–200. https://doi.org/10.1001/archpsyc.64.2.193
Hayashi T, Umemori H, Mishina M, Yamamoto T (1999) The AMPA receptor interacts with and signals through the protein tyrosine kinase Lyn. Nature 397:72–76. https://doi.org/10.1038/16269
Holtzheimer PE 3rd, Mayberg HS (2010) Deep brain stimulation for treatment-resistant depression. Am J Psychiatry 167:1437–1444. https://doi.org/10.1176/appi.ajp.2010.10010141
Ionescu DF, Rosenbaum JF, Alpert JE (2015) Pharmacological approaches to the challenge of treatment-resistant depression. Dialogues Clin Neurosci 17:111–126. https://doi.org/10.31887/DCNS.2015.17.2/dionescu

Jackson AC, Nicoll RA (2011) Stargazin (TARP γ-2) is required for compartment-specific AMPA receptor trafficking and synaptic plasticity in cerebellar stellate cells. J Neurosci 31:3939–3952. https://doi.org/10.1523/JNEUROSCI.5134-10.2011
Jacobi E, von Engelhardt J (2017) Diversity in AMPA receptor complexes in the brain. Curr Opin Neurobiol 45:32–38. https://doi.org/10.1016/j.conb.2017.03.001
Jedema HP, Moghaddam B (1994) Glutamatergic control of dopamine release during stress in the rat prefrontal cortex. J Neurochem 63:785–788. https://doi.org/10.1046/j.1471-4159.1994.63020785.x
Jeon SH, Seong YS, Juhnn YS, Kang UG, Ha KS, Kim YS, Park JB (1997) Electroconvulsive shock increases the phosphorylation of cyclic AMP response element binding protein at Ser-133 in rat hippocampus but not in cerebellum. Neuropharmacology 36:411–414. https://doi.org/10.1016/s0028-3908(97)00047-6
Jiménez-Sánchez L, Castañé A, Pérez-Caballero L, Grifoll-Escoda M, López-Gil X, Campa L, Galofré M, Berrocoso E, Adell A (2016) Activation of AMPA receptors mediates the antidepressant action of deep brain stimulation of the infralimbic prefrontal cortex. Cereb Cortex 26:2778–2789. https://doi.org/10.1093/cercor/bhv133
Jourdi H, Hsu YT, Zhou M, Qin Q, Bi X, Baudry M (2009) Positive AMPA receptor modulation rapidly stimulates BDNF release and increases dendritic mRNA translation. J Neurosci 29:8688–8697. https://doi.org/10.1523/JNEUROSCI.6078-08.2009
Kallarackal AJ, Kvarta MD, Cammarata E, Jaberi L, Cai X, Bailey AM, Thompson SM (2013) Chronic stress induces a selective decrease in AMPA receptor-mediated synaptic excitation at hippocampal temporoammonic-CA1 synapses. J Neurosci 33:15669–15674. https://doi.org/10.1523/JNEUROSCI.2588-13.2013
Karasawa J, Shimazaki T, Kawashima N, Chaki S (2005) AMPA receptor stimulation mediates the antidepressant-like effect of a group II metabotropic glutamate receptor antagonist. Brain Res 1042:92–98. https://doi.org/10.1016/j.brainres.2005.02.032
Kennedy SH, Giacobbe P, Rizvi SJ, Placenza FM, Nishikawa Y, Mayberg HS, Lozano AM (2011) Deep brain stimulation for treatment-resistant depression: follow-up after 3 to 6 years. Am J Psychiatry 168:502–510. https://doi.org/10.1176/appi.ajp.2010.10081187
Kim YK, Lee HP, Won SD, Park EY, Lee HY, Lee BH, Lee SW, Yoon D, Han C, Kim DJ, Choi SH (2007) Low plasma BDNF is associated with suicidal behavior in major depression. Prog Neuro-Psychopharmacol Biol Psychiatry 31:78–85. https://doi.org/10.1016/j.pnpbp.2006.06.024
Kiselycznyk C, Jury NJ, Halladay LR, Nakazawa K, Mishina M, Sprengel R, Grant SG, Svenningsson P, Holmes A (2015) NMDA receptor subunits and associated signaling molecules mediating antidepressant-related effects of NMDA-GluN2B antagonism. Behav Brain Res 287:89–95. https://doi.org/10.1016/j.bbr.2015.03.023
Knapp RJ, Goldenberg R, Shuck C, Cecil A, Watkins J, Miller C, Crites G, Malatynska E (2002) Antidepressant activity of memory-enhancing drugs in the reduction of submissive behavior model. Eur J Pharmacol 440:27–35. https://doi.org/10.1016/s0014-2999(02)01338-9
Koike H, Iijima M, Chaki S (2011) Involvement of AMPA receptor in both the rapid and sustained antidepressant-like effects of ketamine in animal models of depression. Behav Brain Res 224:107–111. https://doi.org/10.1016/j.bbr.2011.05.035
Koike H, Chaki S (2014) Requirement of AMPA receptor stimulation for the sustained antidepressant activity of ketamine and LY341495 during the forced swim test in rats. Behav Brain Res 271:111–115. https://doi.org/10.1016/j.bbr.2014.05.065
Krishnan V, Nestler EJ (2008) The molecular neurobiology of depression. Nature 455:894–902. https://doi.org/10.1038/nature07455
Kunugi A, Tanaka M, Suzuki A, Tajima Y, Suzuki N, Suzuki M, Nakamura S, Kuno H, Yokota A, Sogabe S, Kosugi Y, Awasaki Y, Kaku T, Kimura H (2019) TAK-137, an AMPA-R potentiator with little agonistic effect, has a wide therapeutic window. Neuropsychopharmacology 44:961–970. https://doi.org/10.1038/s41386-018-0213-7
Lapidus KA, Soleimani L, Murrough JW (2013) Novel glutamatergic drugs for the treatment of mood disorders. Neuropsychiatr Dis Treat 9:1101–1112. https://doi.org/10.2147/NDT.S36689

Lauterborn JC, Lynch G, Vanderklish P, Arai A, Gall CM (2000) Positive modulation of AMPA receptors increases neurotrophin expression by hippocampal and cortical neurons. J Neurosci 20:8–21. https://doi.org/10.1523/JNEUROSCI.20-01-00008.2000

Legutko B, Li X, Skolnick P (2001) Regulation of BDNF expression in primary neuron culture by LY392098, a novel AMPA receptor potentiator. Neuropharmacology 40:1019–1027. https://doi.org/10.1016/s0028-3908(01)00006-5

Lener MS, Niciu MJ, Ballard ED, Park M, Park LT, Nugent AC, Zarate CA Jr (2017) Glutamate and gamma-aminobutyric acid systems in the pathophysiology of major depression and antidepressant response to ketamine. Biol Psychiatry 81:886–897. https://doi.org/10.1016/j.biopsych.2016.05.005

Lepack AE, Fuchikami M, Dwyer JM, Banasr M, Duman RS (2014) BDNF release is required for the behavioral actions of ketamine. Int J Neuropsychopharmacol 18:pyu033. https://doi.org/10.1093/ijnp/pyu033

Li N, Lee B, Liu RJ, Banasr M, Dwyer JM, Iwata M, Li XY, Aghajanian G, Duman RS (2010) mTOR-dependent synapse formation underlies the rapid antidepressant effects of NMDA antagonists. Science 329:959–964. https://doi.org/10.1126/science.1190287

Li X, Tizzano JP, Griffey K, Clay M, Lindstrom T, Skolnick P (2001) Antidepressant-like actions of an AMPA receptor potentiator (LY392098). Neuropharmacology 40:1028–1033. https://doi.org/10.1016/s0028-3908(00)00194-5

Li X, Witkin JM, Need AB, Skolnick P (2003) Enhancement of antidepressant potency by a potentiator of AMPA receptors. Cell Mol Neurobiol 23:419–430. https://doi.org/10.1023/a:1023648923447

Lindefors N, Ballarin M, Ernfors P, Falkenberg T, Persson H (1992) Stimulation of glutamate receptors increases expression of brain-derived neurotrophic factor mRNA in rat hippocampus. Ann N Y Acad Sci 648:296–299. https://doi.org/10.1111/j.1749-6632.1992.tb24561.x

Liu SJ, Zukin RS (2007) Ca^{2+}-permeable AMPA receptors in synaptic plasticity and neuronal death. Trends Neurosci 30:126–134. https://doi.org/10.1016/j.tins.2007.01.006

Liu RJ, Lee FS, Li XY, Bambico F, Duman RS, Aghajanian GK (2012) Brain-derived neurotrophic factor Val66Met allele impairs basal and ketamine-stimulated synaptogenesis in prefrontal cortex. Biol Psychiatry 71:996–1005. https://doi.org/10.1016/j.biopsych.2011.09.030

López-Gil X, Jiménez-Sánchez L, Campa L, Castro E, Frago C, Adell A (2019) Role of serotonin and noradrenaline in the rapid antidepressant action of ketamine. ACS Chem Neurosci 10:3318–3326. https://doi.org/10.1021/acschemneuro.9b00288

Lüscher C, Malenka RC (2011) Drug-evoked synaptic plasticity in addiction: from molecular changes to circuit remodeling. Neuron 69:650–663. https://doi.org/10.1016/j.neuron.2011.01.017

Lynch G (2006) Glutamate-based therapeutic approaches: ampakines. Curr Opin Pharmacol 6:82–88. https://doi.org/10.1016/j.coph.2005.09.005

Lynch G, Gall CM (2006) Ampakines and the threefold path to cognitive enhancement. Trends Neurosci 29:554–562. https://doi.org/10.1016/j.tins.2006.07.007

Lynch G (2002) Memory enhancement: the search for mechanism-based drugs. Nat Neurosci 5:1035–1038. https://doi.org/10.1038/nn935

Machado-Vieira R, Salvadore G, Ibrahim I, Diaz-Granados N, Zarate CA (2009) Targeting glutamatergic signaling for the development of novel therapeutics for mood disorders. Curr Pharm Des 15:1595–1611. https://doi.org/10.2174/138161209788168010

Mackowiak M, O'Neill MJ, Hicks CA, Bleakman D, Skolnick P (2002) An AMPA receptor potentiator modulates hippocampal expression of BDNF: an in vivo study. Neuropharmacology 43:1–10. https://doi.org/10.1016/s0028-3908(02)00066-7

Maeng S, Zarate CA Jr, Du J, Schloesser RJ, McCammon J, Chen G, Manji HK (2008) Cellular mechanisms underlying the antidepressant effects of ketamine: role of α-amino-3-hydroxy-5--methylisoxazole-4-propionic acid receptors. Biol Psychiatry 63:349–352. https://doi.org/10.1016/j.biopsych.2007.05.028

Magariños AM, McEwen BS, Flügge G, Fuchs E (1996) Chronic psychosocial stress causes apical dendritic atrophy of hippocampal CA3 pyramidal neurons in subordinate tree shrews. J Neurosci 16:3534–3540. https://doi.org/10.1523/JNEUROSCI.16-10-03534.1996

Malone DA Jr, Dougherty DD, Rezai AR, Carpenter LL, Friehs GM, Eskandar EN, Rauch SL, Rasmussen SA, Machado AG, Kubu S, Tyrka AR, Price LH, Stypulkowski PH, Giftakis JE, Rise MT, Malloy PF, Salloway SP, Greenberg BD (2009) Deep brain stimulation of the ventral capsule/ventral striatum for treatment-resistant depression. Biol Psychiatry 65:267–275. https://doi.org/10.1016/j.biopsych.2008.08.029

Martínez-Turrillas R, Del Río J, Frechilla D (2007) Neuronal proteins involved in synaptic targeting of AMPA receptors in rat hippocampus by antidepressant drugs. Biochem Biophys Res Commun 353:750–755. https://doi.org/10.1016/j.bbrc.2006.12.078

Martínez-Turrillas R, Del Río J, Frechilla D (2005) Sequential changes in BDNF mRNA expression and synaptic levels of AMPA receptor subunits in rat hippocampus after chronic antidepressant treatment. Neuropharmacology 49:1178–1188. https://doi.org/10.1016/j.neuropharm.2005.07.006

Martinez-Turrillas R, Frechilla D, Del Río J (2002) Chronic antidepressant treatment increases the membrane expression of AMPA receptors in rat hippocampus. Neuropharmacology 43:1230–1237. https://doi.org/10.1016/s0028-3908(02)00299-x

Mayberg HS, Lozano AM, Voon V, McNeely HE, Seminowicz D, Hamani C, Schwalb JM, Kennedy SH (2005) Deep brain stimulation for treatment-resistant depression. Neuron 45:651–660. https://doi.org/10.1016/j.neuron.2005.02.014

McEwen BS (2004) Structural plasticity of the adult brain: how animal models help us understand brain changes in depression and systemic disorders related to depression. Dialogues Clin Neurosci 6:119–133. https://doi.org/10.31887/DCNS.2004.6.2/bmcewen

Meador-Woodruff JH, Hogg AJ Jr, Smith RE (2001) Striatal ionotropic glutamate receptor expression in schizophrenia, bipolar disorder, and major depressive disorder. Brain Res Bull 55:631–640. https://doi.org/10.1016/s0361-9230(01)00523-8

Meldrum BS (1994) The role of glutamate in epilepsy and other CNS disorders. Neurology 44(Suppl 8):S14–S23

Mendez-David I, Guilloux JP, Papp M, Tritschler L, Mocaer E, Gardier AM, Bretin S, David DJ (2017) S 47445 produces antidepressant- and anxiolytic-like effects through neurogenesis dependent and independent mechanisms. Front Pharmacol 8:462. https://doi.org/10.3389/fphar.2017.00462

Murrough JW, Abdallah CG, Mathew SJ (2017) Targeting glutamate signalling in depression: progress and prospects. Nat Rev Drug Discov 16:472–486. https://doi.org/10.1038/nrd.2017.16

Musazzi L, Treccani G, Mallei A, Popoli M (2013) The action of antidepressants on the glutamate system: regulation of glutamate release and glutamate receptors. Biol Psychiatry 73:1180–1188. https://doi.org/10.1016/j.biopsych.2012.11.009

Nibuya M, Nestler EJ, Duman RS (1996) Chronic antidepressant administration increases the expression of cAMP response element binding protein (CREB) in rat hippocampus. J Neurosci 16:2365–2372. https://doi.org/10.1523/JNEUROSCI.16-07-02365.1996

Nikonenko I, Boda B, Steen S, Knott G, Welker E, Muller D (2008) PSD-95 promotes synaptogenesis and multiinnervated spine formation through nitric oxide signaling. J Cell Biol 183:1115–1127. https://doi.org/10.1083/jcb.200805132

Noga JT, Wang H (2002) Further postmortem autoradiographic studies of AMPA receptor binding in schizophrenia. Synapse 45:250–258. https://doi.org/10.1002/syn.10106

Oertel BG, Felden L, Tran PV, Bradshaw MH, Angst MS, Schmidt H, Johnson S, Greer JJ, Geisslinger G, Varney MA, Lötsch J (2010) Selective antagonism of opioid-induced ventilatory depression by an ampakine molecule in humans without loss of opioid analgesia. Clin Pharmacol Ther 87:204–211. https://doi.org/10.1038/clpt.2009.194

O'Neill MJ, Witkin JM (2007) AMPA receptor potentiators: application for depression and Parkinson's disease. Curr Drug Targets 8:603–620. https://doi.org/10.2174/138945007780618517

Partin KM (2015) AMPA receptor potentiators: from drug design to cognitive enhancement. Curr Opin Pharmacol 20:46–53. https://doi.org/10.1016/j.coph.2014.11.002

Pascual-Brazo J, Castro E, Díaz A, Valdizán EM, Pilar-Cuéllar F, Vidal R, Treceño B, Pazos A (2012) Modulation of neuroplasticity pathways and antidepressant-like behavioural responses following the short-term (3 and 7 days) administration of the 5-HT_4 receptor agonist RS67333. Int J Neuropsychopharmacol 15:631–643. https://doi.org/10.1017/S1461145711000782

Payne HL (2008) The role of transmembrane AMPA receptor regulatory proteins (TARPs) in neurotransmission and receptor trafficking (review). Mol Membr Biol 25:353–362. https://doi.org/10.1080/09687680801986480

Pilar-Cuellar F, Castro E, Bretin S, Mocaer E, Pazos Á, Díaz Á (2019) S 47445 counteracts the behavioral manifestations and hippocampal neuroplasticity changes in bulbectomized mice. Prog Neuro-Psychopharmacol Biol Psychiatry 93:205–213. https://doi.org/10.1016/j.pnpbp.2019.04.005

Pittenger C, Duman RS (2008) Stress, depression, and neuroplasticity: a convergence of mechanisms. Neuropsychopharmacology 33:88–109. https://doi.org/10.1038/sj.npp.1301574

Pliakas AM, Carlson RR, Neve RL, Konradi C, Nestler EJ, Carlezon WA Jr (2001) Altered responsiveness to cocaine and increased immobility in the forced swim test associated with elevated cAMP response element-binding protein expression in nucleus accumbens. J Neurosci 21:7397–7403. https://doi.org/10.1523/JNEUROSCI.21-18-07397.2001

Price JL, Drevets WC (2012) Neural circuits underlying the pathophysiology of mood disorders. Trends Cogn Sci 16:61–71. https://doi.org/10.1016/j.tics.2011.12.011

Puigdemont D, Pérez-Egea R, Portella MJ, Molet J, de Diego-Adeliño J, Gironell A, Radua J, Gómez-Anson B, Rodríguez R, Serra M, de Quintana C, Artigas F, Álvarez E, Pérez V (2012) Deep brain stimulation of the subcallosal cingulate gyrus: further evidence in treatment-resistant major depression. Int J Neuropsychopharmacol 15:121–133. https://doi.org/10.1017/S1461145711001088

Qin Y, Karst H, Joëls M (2004) Chronic unpredictable stress alters gene expression in rat single dentate granule cells. J Neurochem 89:364–374. https://doi.org/10.1111/j.1471-4159.2003.02332.x

Radin DP, Johnson S, Purcell R, Lippa AS (2018) Effects of chronic systemic low-impact ampakine treatment on neurotrophin expression in rat brain. Biomed Pharmacother 105:540–544. https://doi.org/10.1016/j.biopha.2018.06.008

Ranganathan M, DeMartinis N, Huguenel B, Gaudreault F, Bednar MM, Shaffer CL, Gupta S, Cahill J, Sherif MA, Mancuso J, Zumpano L, D'Souza DC (2017) Attenuation of ketamine-induced impairment in verbal learning and memory in healthy volunteers by the AMPA receptor potentiator PF-04958242. Mol Psychiatry 22:1633–1640. https://doi.org/10.1038/mp.2017.6

Ren J, Ding X, Greer JJ (2012) Respiratory depression in rats induced by alcohol and barbiturate and rescue by ampakine CX717. J Appl Physiol 113:1004–1011. https://doi.org/10.1152/japplphysiol.00752.2012

Réus GZ, Stringari RB, Ribeiro KF, Ferraro AK, Vitto MF, Cesconetto P, Souza CT, Quevedo J (2011) Ketamine plus imipramine treatment induces antidepressant-like behavior and increases CREB and BDNF protein levels and PKA and PKC phosphorylation in rat brain. Behav Brain Res 221:166–171. https://doi.org/10.1016/j.bbr.2011.02.024

Roberts BM, Holden DE, Shaffer CL, Seymour PA, Menniti FS, Schmidt CJ, Williams GV, Castner SA (2010) Prevention of ketamine-induced working memory impairments by AMPA potentiators in a nonhuman primate model of cognitive dysfunction. Behav Brain Res 212:41–48. https://doi.org/10.1016/j.bbr.2010.03.039

Rosa ML, Guimarães FS, Pearson RC, Del Bel EA (2002) Effects of single or repeated restraint stress on GluR1 and GluR2 flip and flop mRNA expression in the hippocampal formation. Brain Res Bull 59:117–124. https://doi.org/10.1016/s0361-9230(02)00859-6

Rueter SM, Burns CM, Coode SA, Mookherjee P, Emeson RB (1995) receptor RNA editing in vitro by enzymatic conversion of adenosine to inosine. Science 267:1491–1494. https://doi.org/10.1126/science.7878468

Sanacora G, Zarate CA, Krystal JH, Manji HK (2008) Targeting the glutamatergic system to develop novel, improved therapeutics for mood disorders. Nat Rev Drug Discov 7:426–437. https://doi.org/10.1038/nrd2462

Sanacora G, Treccani G, Popoli M (2012) Towards a glutamate hypothesis of depression: an emerging frontier of neuropsychopharmacology for mood disorders. Neuropharmacology 62:63–77. https://doi.org/10.1016/j.neuropharm.2011.07.036

Sanacora G, Gueorguieva R, Epperson CN, Wu YT, Appel M, Rothman DL, Krystal JH, Mason GF (2004) Subtype-specific alterations of γ-aminobutyric acid and glutamate in patients with major depression. Arch Gen Psychiatry 61:705–713. https://doi.org/10.1001/archpsyc.61.7.705

Santarelli L, Saxe M, Gross C, Surget A, Battaglia F, Dulawa S, Weisstaub N, Lee J, Duman R, Arancio O, Belzung C, Hen R (2003) Requirement of hippocampal neurogenesis for the behavioral effects of antidepressants. Science 301:805–809. https://doi.org/10.1126/science.1083328

Schwendt M, Jezová D (2000) Gene expression of two glutamate receptor subunits in response to repeated stress exposure in rat hippocampus. Cell Mol Neurobiol 20:319–329. https://doi.org/10.1023/a:1007062109386

Sheline YI, Wang PW, Gado MH, Csernansky JG, Vannier MW (1996) Hippocampal atrophy in recurrent major depression. Proc Natl Acad Sci U S A 93:3908–3913. https://doi.org/10.1073/pnas.93.9.3908

Shen M, Lv D, Li S, Zhang Y, Wang Z, Zhao C, Chen X, Wang C (2019) Positive allosteric modulation of AMPAR by PF-4778574 produced rapid onset antidepressant actions in mice. Cereb Cortex 29:4438–4451. https://doi.org/10.1093/cercor/bhy324

Singh JB, Fedgchin M, Daly EJ, De Boer P, Cooper K, Lim P, Pinter C, Murrough JW, Sanacora G, Shelton RC, Kurian B, Winokur A, Fava M, Manji H, Drevets WC, Van Nueten L (2016) A double-blind, randomized, placebo-controlled, dose-frequency study of intravenous ketamine in patients with treatment-resistant depression. Am J Psychiatry 173:816–826. https://doi.org/10.1176/appi.ajp.2016.16010037

Skolnick P, Legutko B, Li X, Bymaster FP (2001) Current perspectives on the development of non-biogenic amine-based antidepressants. Pharmacol Res 43:411–423. https://doi.org/10.1006/phrs.2000.0806

Sommer B, Keinänen K, Verdoorn TA, Wisden W, Burnashev N, Herb A, Köhler M, Takagi T, Sakmann B, Seeburg PH (1990) Flip and flop: a cell-specific functional switch in glutamate-operated channels of the CNS. Science 249:1580–1585. https://doi.org/10.1126/science.1699275

Song I, Huganir RL (2002) Regulation of AMPA receptors during synaptic plasticity. Trends Neurosci 25:578–588. https://doi.org/10.1016/s0166-2236(02)02270-1

Spellman T, Liston C (2020) Toward circuit mechanisms of pathophysiology in depression. Am J Psychiatry 177:381–390. https://doi.org/10.1176/appi.ajp.2020.20030280

Straub C, Tomita S (2012) The regulation of glutamate receptor trafficking and function by TARPs and other transmembrane auxiliary subunits. Curr Opin Neurobiol 22:488–495. https://doi.org/10.1016/j.conb.2011.09.005

Suenaga T, Morinobu S, Kawano K, Sawada T, Yamawaki S (2004) Influence of immobilization stress on the levels of CaMKII and phospho-CaMKII in the rat hippocampus. Int J Neuropsychopharmacol 7:299–309. https://doi.org/10.1017/S1461145704004304

Svenningsson P, Tzavara ET, Witkin JM, Fienberg AA, Nomikos GG, Greengard P (2002) Involvement of striatal and extrastriatal DARPP-32 in biochemical and behavioral effects of fluoxetine (Prozac). Proc Natl Acad Sci U S A 99:3182–3187. https://doi.org/10.1073/pnas.052712799

Svenningsson P, Kim Y, Warner-Schmidt J, Oh YS, Greengard P (2013) p11 and its role in depression and therapeutic responses to antidepressants. Nat Rev Neurosci 14:673–680. https://doi.org/10.1038/nrn3564

Svenningsson P, Chergui K, Rachleff I, Flajolet M, Zhang X, El Yacoubi M, Vaugeois JM, Nomikos GG, Greengard P (2006) Alterations in 5-HT1B receptor function by p11 in depression-like states. Science 311:77–80. https://doi.org/10.1126/science.1117571

Tan CH, He X, Yang J, Ong WY (2006) Changes in AMPA subunit expression in the mouse brain after chronic treatment with the antidepressant maprotiline: a link between noradrenergic and glutamatergic function? Exp Brain Res 170:448–456. https://doi.org/10.1007/s00221-005-0228-2

Thome J, Sakai N, Shin K, Steffen C, Zhang YJ, Impey S, Storm D, Duman RS (2000) cAMP response element-mediated gene transcription is upregulated by chronic antidepressant treatment. J Neurosci 20:4030–4036. https://doi.org/10.1523/JNEUROSCI.20-11-04030.2000

Toth E, Gersner R, Wilf-Yarkoni A, Raizel H, Dar DE, Richter-Levin G, Levit O, Zangen A (2008) Age-dependent effects of chronic stress on brain plasticity and depressive behavior. J Neurochem 107:522–532. https://doi.org/10.1111/j.1471-4159.2008.05642.x

Traynelis SF, Wollmuth LP, McBain CJ, Menniti FS, Vance KM, Ogden KK, Hansen KB, Yuan H, Myers SJ, Dingledine R (2010) Glutamate receptor ion channels: structure, regulation, and function. Pharmacol Rev 62:405–496. https://doi.org/10.1124/pr.109.002451

Vaidya VA, Duman RS (2001) Depression-emerging insights from neurobiology. Br Med Bull 57:61–79. https://doi.org/10.1093/bmb/57.1.61

Warner-Schmidt JL, Schmidt EF, Marshall JJ, Rubin AJ, Arango-Lievano M, Kaplitt MG, Ibañez-Tallon I, Heintz N, Greengard P (2012) Cholinergic interneurons in the nucleus accumbens regulate depression-like behavior. Proc Natl Acad Sci U S A 109:11360–11365. https://doi.org/10.1073/pnas.1209293109

Warner-Schmidt JL, Chen EY, Zhang X, Marshall JJ, Morozov A, Svenningsson P, Greengard P (2010) A role for p11 in the antidepressant action of brain-derived neurotrophic factor. Biol Psychiatry 68:528–535. https://doi.org/10.1016/j.biopsych.2010.04.029

Watanabe Y, Weiland NG, McEwen BS (1995) Effects of adrenal steroid manipulations and repeated restraint stress on dynorphin mRNA levels and excitatory amino acid receptor binding in hippocampus. Brain Res 680(1–2):217–225. https://doi.org/10.1016/0006-8993(95)00235-i

Wesensten NJ, Reichardt RM, Balkin TJ (2007) Ampakine (CX717) effects on performance and alertness during simulated night shift work. Aviat Space Environ Med 78:937–943. https://doi.org/10.3357/asem.2055.2007

Xing J, Ginty DD, Greenberg ME (1996) Coupling of the RAS-MAPK pathway to gene activation by RSK2, a growth factor-regulated CREB kinase. Science 273:959–963. https://doi.org/10.1126/science.273.5277.959

Yuen EY, Wei J, Liu W, Zhong P, Li X, Yan Z (2012) Repeated stress causes cognitive impairment by suppressing glutamate receptor expression and function in prefrontal cortex. Neuron 73:962–977. https://doi.org/10.1016/j.neuron.2011.12.033

Zafra F, Hengerer B, Leibrock J, Thoenen H, Lindholm D (1990) Activity dependent regulation of BDNF and NGF mRNAs in the rat hippocampus is mediated by non-NMDA glutamate receptors. EMBO J 9:3545–3550

Zanos P, Moaddel R, Morris PJ, Georgiou P, Fischell J, Elmer GI, Alkondon M, Yuan P, Pribut HJ, Singh NS, Dossou KS, Fang Y, Huang XP, Mayo CL, Wainer IW, Albuquerque EX, Thompson SM, Thomas CJ, Zarate CA Jr, Gould TD (2016) NMDAR inhibition-independent antidepressant actions of ketamine metabolites. Nature 533:481–486. https://doi.org/10.1038/nature17998

Zarate CA Jr, Singh JB, Carlson PJ, Brutsche NE, Ameli R, Luckenbaugh DA, Charney DS, Manji HK (2006) A randomized trial of an N-methyl-D-aspartate antagonist in treatment-resistant major depression. Arch Gen Psychiatry 63:856–864. https://doi.org/10.1001/archpsyc.63.8.856

Zarate C Jr, Machado-Vieira R, Henter I, Ibrahim L, Diazgranados N, Salvadore G (2010) Glutamatergic modulators: the future of treating mood disorders? Harv Rev Psychiatry 18:293–303. https://doi.org/10.3109/10673229.2010.511059

Zheng Y, Balabhadrapatruni S, Masumura C, Darlington CL, Smith PF (2011) Effects of the putative cognitive-enhancing ampakine, CX717, on attention and object recognition memory. Curr Alzheimer Res 8:876–882. https://doi.org/10.2174/156720511798192709

Zhou W, Wang N, Yang C, Li XM, Zhou ZQ, Yang JJ (2014) Ketamine-induced antidepressant effects are associated with AMPA receptors-mediated upregulation of mTOR and BDNF in rat hippocampus and prefrontal cortex. Eur Psychiatry 29:419–423. https://doi.org/10.1016/j.eurpsy.2013.10.005

Zwart R, Sher E, Ping X, Jin X, Sims JR Jr, Chappell AS, Gleason SD, Hahn PJ, Gardinier K, Gernert DL, Hobbs J, Smith JL, Valli SN, Witkin JM (2014) Perampanel, an antagonist of α-amino-3-hydroxy-5-methyl-4-isoxazolepropionic acid receptors, for the treatment of epilepsy: studies in human epileptic brain and nonepileptic brain and in rodent models. J Pharmacol Exp Ther 351:124–133. https://doi.org/10.1124/jpet.114.212779

mGlu2/3 Receptor Antagonists as Rapid-Acting Antidepressants

Shigeyuki Chaki

Abstract Since the discovery of the robust and rapid-acting antidepressant effects of ketamine in patients with treatment-resistant depression, the glutamatergic system has gained significant attention as an attractive target for the development of novel antidepressants. As part of the glutamatergic system, metabotropic glutamate (mGlu) receptors are of interest because mGlu receptors play regulatory roles in glutamate transmission. Accumulating evidence has suggested that a blockade of group II mGlu receptors, consisting of mGlu2 and mGlu3 receptors, is an effective approach to exert antidepressant actions. Indeed, animal studies using selective mGlu2/3 receptor antagonists have revealed that mGlu2/3 receptor antagonists have antidepressant profiles similar to those of ketamine: rapid-acting antidepressant effects and efficacy in models refractory to current medications. Moreover, mGlu2/3 receptor antagonists do not show ketamine-like side effects such as psychotomimetic-like behaviors and abuse potential. Therefore, mGlu2/3 receptor antagonists can function as novel antidepressants that may be devoid of ketamine-like side effects.

Keywords mGlu2/3 receptor antagonist · Ketamine · Antidepressant · Brain-derived neurotrophic factor · AMPA receptor

Introduction

Abnormalities of the glutamatergic system have been implicated in the pathophysiology of depression. Indeed, the glutamatergic system has recently gained significant attention as a target for drug discovery for novel antidepressants (Chaki 2017a). Glutamate exerts its effects via two major classes of receptors, ionotropic glutamate (iGlu) receptors and metabotropic glutamate (mGlu) receptors. iGlu receptors are divided into three receptor types (α-amino-3-hydroxy-5-methyl-4-isoxazolepropionic acid [AMPA], kainate, and *N*-methyl-D-aspartate [NMDA]),

S. Chaki (✉)
Research Headquarters, Taisho Pharmaceutical Co., Ltd., Saitama, Saitama, Japan
e-mail: s-chaki@taisho.co.jp

K. Hashimoto, M. Manto (eds.), *New Rapid-acting Antidepressants*, Contemporary Clinical Neuroscience,
https://doi.org/10.1007/978-3-030-79790-4_7

among which both NMDA receptors and AMPA receptors have been investigated as attractive targets for antidepressants. This assumption is stemmed from studies on the mechanisms of ketamine, which has been demonstrated to display rapid-acting antidepressant actions in patients with treatment-resistant depression (TRD) (Berman et al. 2000; Krystal et al. 2013; Monteggia and Zarate 2015; Zarate et al. 2006). mGlu receptors, which are class C G-protein-coupled receptors, consist of eight subtypes that are classified into three major groups (Groups I–III), based on amino acid homology, pharmacological characteristics, and signal transduction mechanisms (Conn and Pin 1997; Schoepp and Conn 1993). Group I mGlu receptors include the mGlu1 receptor and the mGlu5 receptor, which are coupled to Gq/G11 to activate phospholipase C. In addition, postsynaptic mGlu5 receptors are functionally linked to NMDA receptors to positively modulate their activity. By contrast, both Group II mGlu receptors (mGlu2 and mGlu3 receptors) and Group III mGlu receptors (mGlu4, mGlu6, mGlu7, and mGlu8 receptors) are coupled to Gi/Go to negatively modulate adenylyl cyclase activity. Because of their modulatory roles in glutamatergic transmissions, much attention has been paid to mGlu receptors as drug targets for psychiatric disorders including depression (Chaki et al. 2013, 2019; Pilc et al. 2008). Among glutamate receptors, this chapter focuses on Group II mGlu receptors. For the potential of other mGlu receptors as drug targets for novel antidepressants, please refer to previous reviews (Chaki et al. 2013, 2019; Pilc et al. 2008).

mGlu2/3 Receptors: Molecular Properties, Functions, and Roles in Psychiatric Disorders

Group II mGlu receptors, which consist of mGlu2 and mGlu3 receptors, are highly expressed within the cortical and limbic regions of the brain (Wright et al. 2013). The mGlu2 receptor is mainly localized at the perisynaptic position and functions as an autoreceptor or heteroreceptor (Ohishi et al. 1993a; Neki et al. 1996), while the mGlu3 receptor is localized at both presynaptic and postsynaptic sites and on glial cells (Ohishi et al. 1993b; Tanabe et al. 1993). The presynaptic mGlu2/3 receptors negatively regulate the release of neurotransmitters, including glutamate and GABA (Hiyoshi et al. 2014; Neale and Salt 2006). The functions of the mGlu3 receptor expressed on the glial cells are not yet fully known. However, the mGlu3 receptor on astrocytes may be involved in neuroprotective actions (Durand et al. 2013) while also increasing the release of the inflammatory cytokine interleukin-6 in the presence of interleukin-1β (Aronica et al. 2005).

To date, several ligands acting on mGlu2 and/or mGlu3 receptors, including agonists, antagonists, positive allosteric modulators (PAMs), and negative allosteric modulators (NAMs), have been synthesized. Representative ligands are illustrated in Fig. 1. Roles of the mGlu2/3 receptors in psychiatric disorders have been investigated using these mGlu2/3 receptor ligands, and their therapeutic potential as novel antipsychotics has been noted. Indeed, both mGlu2/3 receptor orthosteric agonists and mGlu2 PAMs have demonstrated efficacy in a variety of animal models of schizophrenia (Chaki 2010), and some of the compounds have been tested in clinical trials, particularly LY2140023, a prodrug of LY404039. Although the clinical trials of LY2140023 were terminated because it failed to demonstrate efficacy, the post hoc analysis supports the concept that stimulation of the mGlu2/3 receptor may ameliorate hyperglutamatergic states, which is observed in early psychosis (Kinon et al. 2015). In addition, the potential of mGlu2/3 receptor agonists for the treatment anxiety disorders was assessed, and mGlu2/3 receptor agonists were effective for generalized anxiety disorders (Dunayevich et al. 2008). Changes in the expression levels of mGlu2/3 receptors were reported in certain regions of postmortem brains of patients with schizophrenia or depression (Feyissa et al. 2010; Ghose et al. 2009). All these results clearly indicate that the mGlu2 receptor has important roles in the pathophysiology of psychiatric disorders.

Antidepressant Profiles of mGlu2/3 Receptor Antagonists in Rodent Models

To date, the antidepressant effects of mGlu2/3 receptor antagonists have been demonstrated in several animal models (Chaki et al. 2004; Chaki 2017b). Importantly, the antidepressant profiles of mGlu2/3 receptor antagonists mimicked those of ketamine with regard to their time course and efficacy. Indeed, as reported in preclinical studies of ketamine, mGlu2/3 receptor antagonists reversed depressive-like behaviors induced by either chronic social defeat stress (Dong et al. 2017) or chronic unpredictable stress (Dwyer et al. 2013; Li et al. 2011) 1 day after administration. In these models, conventional antidepressants take several weeks to show their antidepressant effects. Therefore, mGlu2/3 receptor antagonists exert rapid-acting antidepressant effects in rodent models similar to ketamine. Notably, the mGlu2/3 receptor antagonists not only exerted rapid antidepressant effects, but they also produced sustained antidepressant effects that lasted for at least a week after a single administration (Dong et al. 2017; Dwyer et al. 2013), again mirroring the nature of the antidepressant effects of ketamine. Furthermore, reasonable amounts of the mGlu2/3 receptor antagonists such as MGS0039 and LY341495 used in the above-mentioned studies were not detected in the brain at 24 h after administration, as was ketamine (Nakazato et al. 2004). Therefore, the antidepressant effects of both mGlu2/3 receptor antagonists and ketamine, even at 24 h after administration, may be attributed to changes in synaptic plasticity, which will be discussed in section "Synaptic Mechanisms of the Antidepressant Effects of mGlu2/3 Receptor Antagonists".

Moreover, mGlu2/3 receptor antagonists reportedly exerted effects in models in which conventional antidepressants failed to exert their effects. For example, depressive-like behavior (increased immobility in a forced swimming test) induced

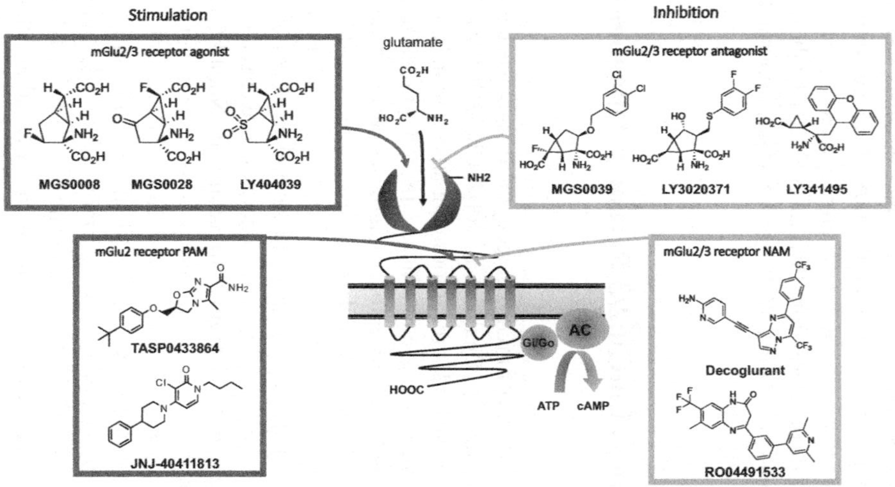

Fig. 1 Representative ligands for mGlu2/3 receptors. To date, numerous ligands (orthosteric agonists, orthosteric antagonists, positive allosteric modulators [PAMs], negative allosteric modulators [NAMs]) for mGlu2/3 receptor have been developed. All of orthosteric agonists and antagonists for mGlu2/3 receptor have glutamate moiety in its molecule, and bind to the orthosteric sites on the N-terminus of the receptor. Both PAMs and NAMs bind to allosteric sites which are located in the transmembrane (TM) domain, TM3, TM5 and TM6. mGlu2 receptor PAMs increase affinity (or potency) for glutamate to the receptor, whereas the mGlu2 receptor PAMs themselves do not activate the receptor

by repeated administration of corticosterone in rats and mice was not reversed by selective serotonin reuptake inhibitors (SSRIs) or imipramine (Ago et al. 2013; Iijima et al. 2010); however, it was reversed by a single administration of ketamine (Koike et al. 2013b). In these models, the depressive-like behavior was also improved by mGlu2/3 receptor antagonists such as MGS0039 and LY341495 (Ago et al. 2013; Koike et al. 2013b). Moreover, learned helplessness behavior (increased escape failures) induced by exposure to certain intensities of inescapable shocks was not reversed by conventional antidepressants even after repeated treatment (Yoshimizu et al. 2006). In the same conditions, both MGS0039 and ketamine significantly reversed the depressive-like behavior in the learned helplessness model (Koike et al. 2011b; Yoshimizu et al. 2006). It has additionally been reported that both mGlu2/3 receptor antagonist and ketamine exerted antidepressant effects in the tail suspension test using CD-1 stain mice, which are relatively resistant to the SSRI citalopram (Witkin et al. 2016). All these findings from preclinical studies indicate that mGlu2/3 receptor antagonists exert antidepressant profiles similar to those of ketamine, including rapid-acting antidepressant effects and efficacy for treatment-resistant models. The rationale that mGlu2/3 receptor blockade displays the antidepressant effects was underpinned by studies in which increased mGlu2/3 receptors were observed in the prefrontal cortex (PFC) in a rodent depression model (Kawasaki et al. 2011) and in depressed patients (Feyissa et al. 2010).

Recently, the role of the mGlu2 receptor in the actions of ketamine has been reported. The antidepressant effects of ketamine were attenuated by an mGlu2/3 receptor agonist (Witkin et al. 2017a; Zanos et al. 2019) and were no longer observed in mice lacking the mGlu2 receptor (Zanos et al. 2019). Notably, the antidepressant effects and the increased γ-power exerted by (2R,6R)-hydroxynorketamine (HNK), an active metabolite of ketamine that has been claimed to be responsible for the antidepressant actions of ketamine (Zanos et al. 2016), were also prevented by mGlu2/3 receptor stimulation. These results further highlight the importance of the mGlu2/3 receptor in antidepressant actions and strengthen the assumption that mGlu2/3 receptor antagonists may have antidepressant profiles similar to that of ketamine.

In addition to ameliorating depressive-like behavior, mGlu2/3 receptor antagonists have been reported to have prophylactic effects against the induction of depressive-like behavior (Highland et al. 2019). Thus, pretreatment with LY341495 prevented escape deficits induced by inescapable shocks or corticosterone treatment, whereas the mGlu2/3 receptor agonist augmented the escape deficits. Notably, the prophylactic effects of LY341495 are observed 24 h after the treatment, at the time when LY341495 does not exist in the brain, indicating the role of synaptic plasticity in its actions. Moreover, knockout mice lacking the mGlu2 receptor were resilient to developing corticosterone-induced escape deficits and chronic social defeat stress-induced anhedonia. Therefore, blockade of the mGlu2/3 receptor may enhance the mechanisms that give resilience to stress. Of note, contrary to the abovementioned results, there have been reports that increased expression or activity of the mGlu2/3 receptor prevents stress-induced abnormalities, as exemplified by the results that mGlu2/3 agonists prevented stress-induced norepinephrine release in the medial prefrontal cortex (mPFC) (Swanson et al. 2004).

mGlu2/3 receptor antagonists have been reported to possess additional characteristics that should allow them to be beneficial as antidepressants. First, the combination of sub-effective doses of LY341495 and ketamine or LY341495 and scopolamine exerted significant antidepressant effects (Podkowa et al. 2016a; Podkowa et al. 2016b; Zanos et al. 2019), indicating that mGlu2/3 receptor antagonists augment the antidepressant effects of ketamine or scopolamine, and the combination of these drugs has synergistic effects. Both ketamine and scopolamine have been known to exert rapid-onset antidepressant effects in patients with major depressive disorder (MDD) including those with TRD (Drevets et al. 2013; Monteggia and Zarate 2015); however, their usages are limited because of side effects. Thus, the adjunctive use of mGlu2/3 receptor antagonists may be useful to mitigate the side effects of ketamine or scopolamine by allowing their doses to be reduced. Secondly, mGlu2/3 receptor antagonists have been reported to show anxiolytic (Shimazaki et al. 2004; Yoshimizu et al. 2006) and pro-cognitive (Goeldner et al. 2013; Shimazaki et al. 2007) effects in some animal models. Both anxiety and cognitive deficits are commonly encountered in patients with depression and are associated with greater symptom severity, impairment, and dysfunction in these patients. Therefore, efficacy against anxiety symptoms is potentially beneficial for depression therapy. Proposed advantages of mGlu2/3 receptor antagonists as antidepressants are summarized in Table 1.

The role of each mGlu2 or mGlu3 receptor in antidepressant actions has yet to be fully elucidated, although there is more evidence for critical roles of the mGlu2 receptor than for the mGlu3 receptor regarding actions of the mGlu2/3 receptor antagonists. The antidepressant effects of LY341495 and LY3020371 were no longer observed in knockout mice that lacked mGlu2 but expressed mGlu3 (Gleason et al. 2013; Witkin et al. 2017a). Thus, the mGlu2 receptor blockade plays a more important role than the mGlu3 receptor blockade in the antidepressant effects of the

Table 1 Beneficial profiles of mGlu2/3 receptor antagonists as antidepressants

Characteristics	Models tested	References
Rapid-acting effects	Chronic social defeat stress model Chronic unpredictable stress model	Dong et al. (2017) Dwyer et al. (2013)
Efficacy in treatment-resistant models	Corticosterone treated model Learned helplessness model CD-1 mice (tail suspension)	Ago et al. (2013) Koike et al. (2013b) Yoshimizu et al. (2006) Koike et al. (2011b) Witkin et al. (2016)
Anxiolytic effects	Novelty-suppressed feeding Marble burying behavior, etc.	Fukumoto et al. (2014) Shimazaki et al. (2004)
Pro-cognitive effects	Social recognition test Morris water maze test, etc.	Shimazaki et al. (2007) Higgins et al. (2004)
Enhancement of effects of ketamine and scopolamine	Forced swimming test Tail suspension test	Podkowa et al. (2016a, b)

mGlu2/3 receptor antagonists. Furthermore, mGlu2 but not mGlu3 receptors have been reported to promote stress resilience and depressive-like behaviors (Zanos et al. 2019). Mice lacking the mGlu2 receptor but not mice lacking the mGlu3 receptor show resilience to increased depressive-like behaviors (increased escape failures or reduced sucrose preference) induced by inescapable shock, corticosterone treatment, or chronic social defeat stress (Highland et al. 2019). Therefore, mGlu2 receptor blockade has roles not only in the antidepressant actions of mGlu2/3 receptor antagonists but also in giving resilience to stress. In addition to the mGlu2 receptor, the role of the mGlu3 receptor in antidepressant actions has recently been reported. A selective mGlu3 receptor NAM was shown to exert both antidepressant and anxiolytic effects in rodent models (Engers et al. 2015). Moreover, when the tail suspension test was performed, the antidepressant effects of a selective mGlu3 receptor NAM were comparable to that of ketamine, whereas a selective mGlu2 receptor NAM failed to show any significant effect at the appropriate exposure level (Engers et al. 2017). Therefore, these studies indicate that mGlu3 receptor blockade may exert antidepressant effects. However, currently, there is no evidence showing that mGlu3 receptor blockade is involved in the antidepressant actions of mGlu2/3 receptor antagonists, and more studies are necessary to address the role of the mGlu3 receptor in the actions of mGlu2/3 receptor antagonists.

Antidepressant Effects of Stimulation of mGlu2/3 and/or mGlu2 Receptors

In contrast to mGlu2 receptor blockades, stimulation of the mGlu2/3 receptors has been reported to exert antidepressant effects. A selective mGlu2 receptor PAM, THIIC, exerted antidepressant effects in three animal models of depression (Fell et al. 2011); however, the effects were no longer observed in knockout mice lacking the mGlu2 receptor. Therefore, mGlu2 receptor stimulation mediates the effects of THIIC. Moreover, an increase in mGlu2 receptor expression may lead to antidepressant effects. L-Acetylcarnitine (LAC) was reported to exhibit an antidepressant effect in genetic (Flinders sensitive line) and environmental (CUS model) animal models faster than the current medications (Nasca et al. 2013). Simultaneously, LAC epigenetically increased mGlu2 receptor expression in the hippocampus, and the antidepressant effect of LAC was partially attenuated in mGlu2 receptor knockout mice or by an mGlu2/3 receptor antagonist. In addition, contrary to what was described in the above section, an mGlu2 receptor blockade may confer susceptibility to stress because knockout mice lacking the mGlu2 receptor are more susceptible to CUS, and a lower level of mGlu2 receptor expression in the hippocampus was observed in stress-susceptible mice than in resilient mice (Nasca et al. 2015). A previous study reported that LY379268, an orthosteric mGlu2/3 receptor agonist, did not exert antidepressant effects on its own; however, it did shorten the onset of the antidepressant effects of clomipramine from 21 days to 14 days in Flinders sensitive line rats (Matrisciano et al. 2008). Notably, evidence for the antidepressant effects of stimulation of the mGlu2 receptor is limited, and other than THIIC, the antidepressant effects of mGlu2 receptor PAMs or mGlu2/3 receptor agonists have not yet been reported.

Mechanisms Underlying the Antidepressant Effects of mGlu2/3 Receptor Antagonists

To date, both synaptic and neural mechanisms underlying the antidepressant effects of mGlu2/3 receptors have been proposed, as depicted in Fig. 2.

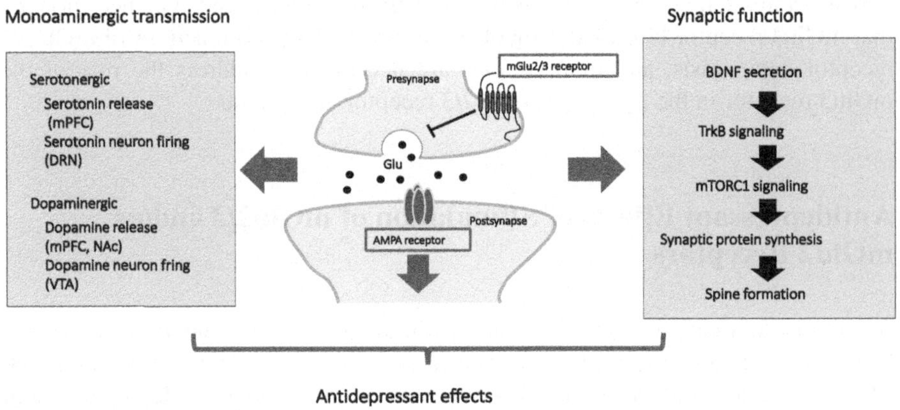

Fig. 2 Proposed synaptic and neuronal mechanisms underlying antidepressant effects of mGlu2/3 receptor antagonists. mGlu2/3 receptor antagonists, by blocking autoreceptor, increases glutamate release, that stimulates postsynaptic AMPA receptor to trigger the subsequent events, on both synaptic and network levels. AMPA receptor stimulation increases activity-dependent BDNF secretion and thereby stimulates TrkB signaling. This signaling cascade induces activation of mTORC1 signaling, resulting in increase in synaptic protein synthesis and spine density. In addition, mGlu2/3 receptor antagonists activates serotonergic system (activation of the DRN serotonin neurons and thereby increase in serotonin release in the mPFC) and dopaminergic system (increase in number of spontaneous firing dopamine neurons in the VTA, and increase in dopamine release in the mPFC and NAc) through AMPA receptor stimulation. All these mechanisms may be involved in the unique antidepressant effects of mGlu2/3 receptor antagonists. BDNF brain-derived neurotrophic factor, TrkB tropomyosin-related kinase B, mTORC1 mechanistic target of rapamycin complex 1, mPFC medial prefrontal cortex, DRN dorsal raphe nucleus, NAc nucleus accumbens, VTA ventral tegmental area

Synaptic Mechanisms of the Antidepressant Effects of mGlu2/3 Receptor Antagonists

Pharmacological studies have revealed that mGlu2/3 receptor antagonists exert the antidepressant effects through activation of the AMPA receptor, tropomyosin-related kinase B (TrkB) signaling, and mechanistic target of rapamycin complex 1 (mTORC1) signaling (Chaki 2017b; Dwyer et al. 2012; Karasawa et al. 2005; Koike et al. 2011a, 2013a). In primary neural cultures, an mGlu2/3 receptor antagonist increased brain-derived neurotrophic factor (BDNF) secretion and phosphorylation of p70S6 kinase, a downstream molecule of mTORC1; increased BDNF secretion was blocked by an AMPA receptor antagonist (Lepack et al. 2016). Therefore, mGlu2/3 receptor antagonists may activate the BDNF/TrkB-mTORC1 signaling cascade via AMPA receptor activation to exert antidepressant effects. It is noteworthy that both mGlu2/3 receptor antagonists and ketamine enhanced AMPA receptor signaling in the hippocampus by metabolomics profiling (Witkin et al. 2017a). Moreover, LY3414945, an mGlu2/3 receptor antagonist, increased the phosphorylation of mTORC1 signaling molecules and increased the levels of synaptic proteins in the mPFC (Dwyer et al. 2012). Likewise, MGS0039 reversed not only the decrease in BDNF expression and phosphorylated TrkB but also reversed the decrease in synaptic protein levels and dendritic spine density induced by chronic social defeat stress in the mPFC and hippocampus (Dong et al. 2017). These results indicate that mGlu2/3 receptor antagonists resume the impaired synaptic dysfunction observed in a depressive state, presumably via mTORC1 signaling. Importantly, these synaptic mechanisms have been proposed as critical mechanisms underlying the antidepressant effects of ketamine (Dong et al. 2017; Krystal et al. 2013; Li et al. 2011). Synaptic dysfunctions including decreased spine density (Duman and Duman 2015) and a reduction in mTORC1 expression and phosphorylation of its signaling molecules (Jernigan et al. 2011) were observed in the PFC of MDD subjects relative to controls. In addition, very recently, it was reported that newly formed dendritic spines are essential for ketamine to exert its antidepressant effects in rodents (Moda-Sava et al. 2019). Therefore, it is convincing to propose that the amelioration of impaired synaptic function by both mGlu2/3 receptor antagonists and ketamine has pivotal roles in their antidepressant actions.

Role of Monoaminergic Systems in the Antidepressant Actions of mGlu2/3 Receptor Antagonists

Currently prescribed antidepressants improve depressive symptoms by globally increasing the concentrations of monoamine neurotransmitters in the brain. Both serotoninergic and dopaminergic systems have also been proposed to mediate the antidepressant effects of the mGlu2/3 receptor antagonists, although their molecular and circuit mechanisms differ from those of conventional antidepressants.

mGlu2/3 receptor antagonists increased the spontaneous firing rates of serotonergic neurons in the dorsal raphe nucleus (DRN) and increased extracellular 5-HT concentrations in the mPFC (Kawashima et al. 2005). The antidepressant effects of LY341495 and ketamine were blocked by the pharmacological depletion of 5-HT and a 5-HT_{1A} receptor antagonist (Fukumoto et al. 2014), suggesting that the serotonergic system, particularly 5-HT_{1A} receptor stimulation, is necessary for mGlu2/3 receptor antagonists and ketamine to exert their antidepressant effects. Interestingly, local injection of the mGlu2/3 receptor antagonist into the mPFC activated the subpopulation of 5-HT neurons in the DRN, the activation of which was caused by AMPA receptor stimulation in the mPFC (Fukumoto et al. 2016). Furthermore, a 5-HT_{1A} receptor antagonist injected into the mPFC prevented the antidepressant effects of LY341495 (Fukumoto et al. 2018a). Likewise, ketamine also exerted the same effects on the serotonergic system via AMPA receptor stimulation in the mPFC, and these mechanisms are necessary for the antidepressant effects of ketamine (Fukumoto et al. 2014, 2016, 2018b). These findings suggest that mGlu2/3 receptor antagonists and ketamine initially activate mPFC neurons in an AMPA receptor-dependent manner, resulting in the activation of DRN serotonergic neurons projecting to the mPFC; the consequent release of 5-HT in the mPFC activates postsynaptic 5-HT_{1A} receptors to exert their antidepressant effects. In concert with our findings, it has been reported that stimulation of the AMPA receptor in the mPFC exerts 5-HT release in the mPFC and antidepressant effects, both through DRN neurons (Gasull-Camós et al. 2017). Moreover, the antidepressant effects induced by the stimulation of the mPFC (infralimbic part) were attenuated by the local 5-HT_{1A} receptor blockade (Gasull-Camós et al. 2018). Therefore, the mPFC 5-HT_{1A} receptor stimulated by the mPFC-DRN projections may be an important neural pathway for both mGlu2/3 receptor antagonists and ketamine to exert their antidepressant effects. This is supported by our results that local 5-HT_{1A} receptor stimulation in the mPFC exerted ketamine-like antidepressant effects, whereas global 5-HT_{1A} receptor stimulation did not (Fukumoto et al. 2018b). By contrast, (*R*)-ketamine, a stereoisomer of racemic ketamine with more potent antidepressant effects than (*S*)-ketamine in rodents (Fukumoto et al. 2017; Yang et al. 2015), exerted its effects in a serotonin-independent manner (Zhang et al. 2018).

In addition to the serotonergic system, mGlu2/3 receptor antagonists and ketamine have been reported to increase the number of spontaneously active dopamine neurons in the ventral tegmental area (VTA) and, accordingly, to elevate extracellular dopamine concentrations in the mPFC and nucleus accumbens, which are dependent on AMPA receptor stimulation (Karasawa et al. 2010; Witkin et al. 2016). Notably, the increase in the number of spontaneously active VTA dopamine neurons was not observed with the SSRI citalopram. However, no direct evidence exists demonstrating that dopaminergic transmission contributes to the antidepressant actions of mGlu2/3 receptor antagonists. Furthermore, although it was reported that the stimulation of dopamine D_1 receptor on the pyramidal neurons in the mPFC was required for the antidepressant effect of ketamine (Hare et al. 2019), but not of (R)-ketamine (Chang et al. 2019), the role of the dopamine D_1 receptor in the effects of mGlu2/3 receptor antagonists has not been demonstrated. Collectively, these findings suggest that both the synaptic mechanisms and the neural mechanisms of mGlu2/3 receptor antagonists are shared with ketamine.

Conclusions

As described above, accumulating evidence from animal studies has indicated that mGlu2/3 receptor antagonists may have antidepressant profiles similar to that of ketamine. In addition, mGlu2/3 receptor antagonists share some mechanisms underlying the antidepressant effects of ketamine, supporting the assumption that mGlu2/3 receptor antagonists may exert ketamine-like antidepressant effects in TRD patients. By contrast, it is noteworthy that mGu2/3 receptor antagonists, unlike ketamine, do not induce psychotomimetic-like behaviors, cognitive impairment, abuse potential, or neurotoxicity (Witkin et al. 2017b), which raises the possibility that mGlu2/3 receptor antagonists may function as antidepressants without ketamine-like side effects. This hypothesis must be proven in clinical studies with patients with TRD. In this regard, the failures of a proof-of-concept study of decoglurant, an mGlu2/3 receptor NAM, are discouraging (Umbricht et al. 2015). However, the validity of the clinical designs has not been fully investigated. It is necessary to carefully investigate whether the profiles of the compound and doses tested were appropriate. In addition, differences between orthosteric antagonists and NAM must be taken into account. Therefore, the efficacy and safety of mGlu2/3 receptor antagonists for patients with TRD must be proven in clinical studies using an appropriate mGlu2/3 receptor antagonist.

References

Ago Y, Yano K, Araki R, Hiramatsu N, Kita Y, Kawasaki T, Onoe H, Chaki S, Nakazato A, Hashimoto H, Baba A, Takuma K, Matsuda T (2013) Metabotropic glutamate 2/3 receptor antagonists improve behavioral and prefrontal dopaminergic alterations in the chronic corticosterone-induced depression model in mice. Neuropharmacology 65:29–38

Aronica E, Gorter JA, Rozemuller AJ, Yankaya B, Troost D (2005) Activation of metabotropic glutamate receptor 3 enhances interleukin (IL)-1beta-stimulated release of IL-6 in cultured human astrocytes. Neuroscience 130:927–933

Berman RM, Cappiello A, Anand A, Oren DA, Heninger GR, Charney DS, Krystal JH (2000) Antidepressant effects of ketamine in depressed patients. Biol Psychiatry 47:351–354

Chaki S (2010) Group II metabotropic glutamate receptor agonists as a potential drug for schizophrenia. Eur J Pharmacol 639:59–66

Chaki S (2017a) Beyond ketamine: new approaches to the development of safer antidepressants. Curr Neuropharmacol 15:963–976

Chaki S (2017b) mGlu2/3 receptor antagonists as novel antidepressants. Trends Pharmacol Sci 38:569–580

Chaki S, Ago Y, Palucha-Paniewiera A, Matrisciano F, Pilc A (2013) mGlu2/3 and mGlu5 receptors: potential targets for novel antidepressants. Neuropharmacology 66:40–52

Chaki S, Koike H, Fukumoto K (2019) Targeting of metabotropic glutamate receptors for the development of novel antidepressants. Chronic Stress 3:1–13

Chaki S, Yoshikawa R, Hirota S, Shimazaki T, Maeda M, Kawashima N, Yoshimizu T, Yasuhara A, Sakagami K, Okuyama S, Nakanishi S, Nakazato A (2004) MGS0039: a potent and selective group II metabotropic glutamate receptor antagonist with antidepressant-like activity. Neuropharmacology 46:457–467

Chang L, Zhang K, Pu Y, Qu Y, Wang SM, Xiong Z, Shirayama Y, Hashimoto K (2019) Lack of dopamine D1 receptors in the antidepressant actions of (R)-ketamine in a chronic social defeat stress model. Eur Arch Psychiatry Clin Neurosci. https://doi.org/10.1007/s00406-019-01012-1

Conn PJ, Pin JP (1997) Pharmacology and functions of metabotropic glutamate receptors. Annu Rev Pharmacol Toxicol 37:205–237

Dong C, Zhang JC, Yao W, Ren Q, Ma M, Yang C, Chaki S, Hashimoto K (2017) Rapid and sustained antidepressant action of the mGlu2/3 receptor antagonist MGS0039 in the social defeat stress model: comparison with ketamine. Int J Neuropsychopharmacol 20:228–236

Drevets WC, Zarate CA Jr, Furey ML (2013) Antidepressant effects of the muscarinic cholinergic receptor antagonist scopolamine: a review. Biol Psychiatry 73:1156–1163

Duman CH, Duman RS (2015) Spine synapse remodeling in the pathophysiology and treatment of depression. Neurosci Lett 601:20–29

Dunayevich E, Erickson J, Levine L, Landbloom R, Schoepp DD, Tollefson GD (2008) Efficacy and tolerability of an mGlu2/3 agonist in the treatment of generalized anxiety disorder. Neuropsychopharmacology 33:1603–1610

Durand D, Carniglia L, Caruso C, Lasaga M (2013) mGlu3 receptor and astrocytes: partners in neuroprotection. Neuropharmacology 66:1–11

Dwyer JM, Lepack AE, Duman RS (2012) mTOR activation is required for the antidepressant effects of mGluR2/3 blockade. Int J Neuropsychopharmacol 15:429–434

Dwyer JM, Lepack AE, Duman RS (2013) mGluR2/3 blockade produces rapid and long-lasting reversal of anhedonia caused by chronic stress exposure. J Mol Psychiatry 1:15

Engers JL, Bollinger KA, Weiner RL, Rodriguez AL, Long MF, Breiner MM, Chang S, Bollinger SR, Bubser M, Jones CK, Morrison RD, Bridges TM, Blobaum AL, Niswender CM, Conn PJ, Emmitte KA, Lindsley CW (2017) Design and synthesis of N-Aryl phenoxyethoxy pyridinones as highly selective and CNS penetrant mGlu3 NAMs. ACS Med Chem Lett 8:925–930

Engers JL, Rodriguez AL, Konkol LC, Morrison RD, Thompson AD, Byers FW, Blobaum AL, Chang S, Venable DF, Loch MT, Niswender CM, Daniels JS, Jones CK, Conn PJ, Lindsley CW, Emmitte KA (2015) Discovery of a selective and CNS penetrant negative allosteric modulator of metabotropic glutamate receptor subtype 3 with antidepressant and anxiolytic activity in rodents. J Med Chem 58:7485–7500

Fell MJ, Witkin JM, Falcone JF, Katner JS, Perry KW, Hart J, Rorick-Kehn L, Overshiner CD, Rasmussen K, Chaney SF, Benvenga MJ, Li X, Marlow DL, Thompson LK, Luecke SK, Wafford KA, Seidel WF, Edgar DM, Quets AT, Felder CC, Wang X, Heinz BA, Nikolayev A, Kuo MS, Mayhugh D, Khilevich A, Zhang D, Ebert PJ, Eckstein JA, Ackermann BL, Swanson SP, Catlow JT, Dean RA, Jackson K, Tauscher-Wisniewski S, Marek GJ, Schkeryantz JM, Svensson KA (2011) N-(4-((2-(trifluoromethyl)-3-hydroxy-4-(isobutyryl)phenoxy)methyl) benzyl)-1-methyl-1H-imidazole-4-carboxamide (THIIC), a novel metabotropic glutamate 2 potentiator with potential anxiolytic/antidepressant properties: in vivo profiling suggests a link between behavioral and central nervous system neurochemical changes. J Pharmacol Exp Ther 336:165–177

Feyissa AM, Woolverton WL, Miguel-Hidalgo JJ, Wang Z, Kyle PB, Hasler G, Stockmeier CA, Iyo AH, Karolewicz B (2010) Elevated level of metabotropic glutamate receptor 2/3 in the prefrontal cortex in major depression. Prog Neuro-Psychopharmacol Biol Psychiatry 34:279–283

Fukumoto K, Iijima M, Chaki S (2014) Serotonin-1A receptor stimulation mediates effects of a metabotropic glutamate 2/3 receptor antagonist, 2S-2-amino-2-(1S,2S-2-carboxycycloprop-1-yl)-3-(xanth-9-yl)propanoic acid (LY341495), and an N-methyl-D-aspartate receptor antagonist, ketamine, in the novelty-suppressed feeding test. Psychopharmacology 231:2291–2298

Fukumoto K, Iijima M, Chaki S (2016) The antidepressant effects of an mGlu2/3 receptor antagonist and ketamine require AMPA receptor stimulation in the mPFC and subsequent activation of the 5-HT neurons in the DRN. Neuropsychopharmacology 41:1046–1056

Fukumoto K, Iijima M, Funakoshi T, Chaki S (2018a) 5-HT1A receptor stimulation in the medial prefrontal cortex mediates the antidepressant effects of mGlu2/3 receptor antagonist in mice. Neuropharmacology 137:96–103

Fukumoto K, Iijima M, Funakoshi T, Chaki S (2018b) Role of 5-HT1A receptor stimulation in the medial prefrontal cortex in the sustained antidepressant effects of ketamine. Int J Neuropsychopharmacol 21:371–381

Fukumoto K, Toki H, Iijima M, Hashihayata T, Yamaguchi JI, Hashimoto K, Chaki S (2017) Antidepressant potential of (R)-ketamine in rodent models: comparison with (S)-ketamine. J Pharmacol Exp Ther 361:9–16

Gasull-Camós J, Martínez-Torres S, Tarrés-Gatius M, Ozaita A, Artigas F, Castañé A (2018) Serotonergic mechanisms involved in antidepressant-like responses evoked by GLT-1 blockade in rat infralimbic cortex. Neuropharmacology 139:41–51

Gasull-Camós J, Tarrés-Gatius M, Artigas F, Castañé A (2017) Glial GLT-1 blockade in infralimbic cortex as a new strategy to evoke rapid antidepressant-like effects in rats. Transl Psychiatry 7:e1038. https://doi.org/10.1038/tp.2017.7

Ghose S, Gleason KA, Potts BW, Lewis-Amezcua K, Tamminga CA (2009) Differential expression of metabotropic glutamate receptors 2 and 3 in schizophrenia: a mechanism for antipsychotic drug action? Am J Psychiatry 166:812–820

Gleason SD, Li X, Smith IA, Ephlin JD, Wang XS, Heinz BA, Carter JH, Baez M, Yu J, Bender DM, Witkin JM (2013) mGlu2/3 agonist-induced hyperthermia: an in vivo assay for detection of mGlu2/3 receptor antagonism and its relation to antidepressant-like efficacy in mice. CNS Neurol Disord Drug Targets 12:554–566

Goeldner C, Ballard TM, Knoflach F, Wichmann J, Gatti S, Umbricht D (2013) Cognitive impairment in major depression and the mGlu2 receptor as a therapeutic target. Neuropharmacology 64:337–346

Hare BD, Shinohara R, Liu RJ, Pothula S, DiLeone RJ, Duman RS (2019) Optogenetic stimulation of medial prefrontal cortex Drd1 neurons produces rapid and long-lasting antidepressant effects. Nat Commun 10:223. https://doi.org/10.1038/s41467-018-08168-9

Highland JN, Zanos P, Georgiou P, Gould TD (2019) Group II metabotropic glutamate receptor blockade promotes stress resilience in mice. Neuropsychopharmacology. https://doi.org/10.1038/s41386-019-0380-1

Higgins GA, Ballard TM, Kew JN, Richards JG, Kemp JA, Adam G, Woltering T, Nakanishi S, Mutel V (2004) Pharmacological manipulation of mGlu2 receptors influences cognitive performance in the rodent. Neuropharmacology 46:907–917

Hiyoshi T, Marumo T, Hikichi H, Tomishima Y, Urabe H, Tamita T, Iida I, Yasuhara A, Karasawa J, Chaki S (2014) Neurophysiologic and antipsychotic profiles of TASP0433864, a novel positive allosteric modulator of metabotropic glutamate 2 receptor. J Pharmacol Exp Ther 351:642–653

Iijima M, Ito A, Kurosu S, Chaki S (2010) Pharmacological characterization of repeated corticosterone injection-induced depression model in rats. Brain Res 1359:75–80

Jernigan CS, Goswami DB, Austin MC, Iyo AH, Chandran A, Stockmeier CA, Karolewicz B (2011) The mTOR signaling pathway in the prefrontal cortex is compromised in major depressive disorder. Prog Neuro-Psychopharmacol Biol Psychiatry 35:1774–1779

Karasawa J, Kotani M, Kambe D, Chaki S (2010) AMPA receptor mediates mGlu 2/3 receptor antagonist-induced dopamine release in the rat nucleus accumbens shell. Neurochem Int 57:615–619

Karasawa J, Shimazaki T, Kawashima N, Chaki S (2005) AMPA receptor stimulation mediates the antidepressant-like effect of a group II metabotropic glutamate receptor antagonist. Brain Res 1042:92–98

Kawasaki T, Ago Y, Yano K, Araki R, Washida Y, Onoe H, Chaki S, Nakazato A, Hashimoto H, Baba A, Takuma K, Matsuda T (2011) Increased binding of cortical and hippocampal group II metabotropic glutamate receptors in isolation-reared mice. Neuropharmacology 60:397–404

Kawashima N, Karasawa J, Shimazaki T, Chaki S, Okuyama S, Yasuhara A, Nakazato A (2005) Neuropharmacological profiles of antagonists of group II metabotropic glutamate receptors. Neurosci Lett 378:131–134

Kinon BJ, Millen BA, Zhang L, McKinzie DL (2015) Exploratory analysis for a targeted patient population responsive to the metabotropic glutamate 2/3 receptor agonist pomaglumetad methionil in schizophrenia. Biol Psychiatry 78:754–762

Koike H, Iijima M, Chaki S (2011a) Involvement of the mammalian target of rapamycin signaling in the antidepressant-like effect of group II metabotropic glutamate receptor antagonists. Neuropharmacology 61:1419–1423

Koike H, Iijima M, Chaki S (2011b) Involvement of AMPA receptor in both the rapid and sustained antidepressant-like effects of ketamine in animal models of depression. Behav Brain Res 224:107–111

Koike H, Iijima M, Chaki S (2013b) Effects of ketamine and LY341495 on the depressive-like behavior of repeated corticosterone-injected rats. Pharmacol Biochem Behav 107:20–23

Koike H, Fukumoto K, Iijima M, Chaki S (2013a) Role of BDNF/TrkB signaling in antidepressant-like effects of a group II metabotropic glutamate receptor antagonist in animal models of depression. Behav Brain Res 238:48–52

Krystal JH, Sanacora G, Duman RS (2013) Rapid-acting glutamatergic antidepressants: the path to ketamine and beyond. Biol Psychiatry 73:1133–1141

Lepack AE, Bang E, Lee B, Dwyer JM, Duman RS (2016) Fast-acting antidepressants rapidly stimulate ERK signaling and BDNF release in primary neuronal cultures. Neuropharmacology 111:242–252

Li N, Liu RJ, Dwyer JM, Banasr M, Lee B, Son H, Li XY, Aghajanian G, Duman RS (2011) Glutamate N-methyl-D-aspartate receptor antagonists rapidly reverse behavioral and synaptic deficits caused by chronic stress exposure. Biol Psychiatry 69:754–761

Matrisciano F, Caruso A, Orlando R, Marchiafava M, Bruno V, Battaglia G, Gruber SH, Melchiorri D, Tatarelli R, Girardi P, Mathè AA, Nicoletti F (2008) Defective group-II metaboropic glutamate receptors in the hippocampus of spontaneously depressed rats. Neuropharmacology 55:525–531

Moda-Sava RN, Murdock MH, Parekh PK, Fetcho RN, Huang BS, Huynh TN, Witztum J, Shaver DC, Rosenthal DL, Alway EJ, Lopez K, Meng Y, Nellissen L, Grosenick L, Milner TA, Deisseroth K, Bito H, Kasai H, Liston C (2019) Sustained rescue of prefrontal circuit dysfunction by antidepressant-induced spine formation. Science 364. pii: eaat8078. https://doi.org/10.1126/science.aat8078

Monteggia LM, Zarate C Jr (2015) Antidepressant actions of ketamine: from molecular mechanisms to clinical practice. Curr Opin Neurobiol 30:139–143

Nakazato A, Sakagami K, Yasuhara A, Ohta H, Yoshikawa R, Itoh M, Nakamura M, Chaki S (2004) Synthesis, in vitro pharmacology, structure-activity relationships, and pharmacokinetics of 3-alkoxy-2-amino-6-fluorobicyclo[3.1.0]hexane-2,6-dicarboxylic acid derivatives as potent and selective group II metabotropic glutamate receptor antagonists. J Med Chem 47:4570–4587

Nasca C, Bigio B, Zelli D, Nicoletti F, McEwen BS (2015) Mind the gap: glucocorticoids modulate hippocampal glutamate tone underlying individual differences in stress susceptibility. Mol Psychiatry 20:755–763

Nasca C, Xenos D, Barone Y, Caruso A, Scaccianoce S, Matrisciano F, Battaglia G, Mathé AA, Pittaluga A, Lionetto L, Simmaco M, Nicoletti F (2013) L-acetylcarnitine causes rapid antidepressant effects through the epigenetic induction of mGlu2 receptors. Proc Natl Acad Sci U S A 110:4804–4809

Neale SA, Salt TE (2006) Modulation of GABAergic inhibition in the rat superior colliculus by a presynaptic group II metabotropic glutamate receptor. J Physiol 577:659–669

Neki A, Ohishi H, Kaneko T, Shigemoto R, Nakanishi S, Mizuno N (1996) Pre- and postsynaptic localization of a metabotropic glutamate receptor, mGluR2, in the rat brain: an immunohistochemical study with a monoclonal antibody. Neurosci Lett 202:197–200
Ohishi H, Shigemoto R, Nakanishi S, Mizuno N (1993a) Distribution of the messenger RNA for a metabotropic glutamate receptor, mGluR2, in the central nervous system of the rat. Neuroscience 53:1009–1018
Ohishi H, Shigemoto R, Nakanishi S, Mizuno N (1993b) Distribution of the mRNA for a metabotropic glutamate receptor (mGluR3) in the rat brain: an in situ hybridization study. J Comp Neurol 335:252–266
Pilc A, Chaki S, Nowak G, Witkin JM (2008) Mood disorders: regulation by metabotropic glutamate receptors. Biochem Pharmacol 75:997–1006
Podkowa K, Pochwat B, Brański P, Pilc A, Pałucha-Poniewiera A (2016a) Group II mGlu receptor antagonist LY341495 enhances the antidepressant-like effects of ketamine in the forced swim test in rats. Psychopharmacology 233:2901–2914
Podkowa K, Podkowa A, Sałat K, Lenda T, Pilc A, Pałucha-Poniewiera A (2016b) Antidepressant-like effects of scopolamine in mice are enhanced by the group II mGlu receptor antagonist LY341495. Neuropharmacology 111:169–179
Schoepp DD, Conn PJ (1993) Metabotropic glutamate receptors in brain function and pathology. Trends Pharmacol Sci 14(1):13–20
Shimazaki T, Iijima M, Chaki S (2004) Anxiolytic-like activity of MGS0039, a potent group II metabotropic glutamate receptor antagonist, in a marble-burying behavior test. Eur J Pharmacol 501:121–125
Shimazaki T, Kaku A, Chaki S (2007) Blockade of the metabotropic glutamate 2/3 receptors enhances social memory via the AMPA receptor in rats. Eur J Pharmacol 575:94–97
Swanson CJ, Perry KW, Schoepp DD (2004) The mGlu2/3 receptor agonist, LY354740, blocks immobilization-induced increases in noradrenaline and dopamine release in the rat medial prefrontal cortex. J Neurochem 88:194–202
Tanabe Y, Nomura A, Masu M, Shigemoto R, Mizuno N, Nakanishi S (1993) Signal transduction, pharmacological properties, and expression patterns of two rat metabotropic glutamate receptors, mGluR3 and mGluR4. J Neurosci 13:1372–1378
Umbricht D et al (2015) Results of a double-blind placebo-controlled study of the antidepressant effects of the mGLU2 negative allosteric modulator RG1578. Eur Neuropsychopharmacol 25:S447
Witkin JM, Mitchell SN, Wafford KA, Carter G, Gilmour G, Li J, Eastwood BJ, Overshiner C, Li X, Rorick-Kehn L, Rasmussen K, Anderson WH, Nikolayev A, Tolstikov VV, Kuo MS, Catlow JT, Li R, Smith SC, Mitch CH, Ornstein PL, Swanson S, Monn JA (2017a) Comparative effects of LY3020371, a potent and selective metabotropic glutamate (mGlu) 2/3 receptor antagonist, and ketamine, a noncompetitive N-methyl-d-aspartate receptor antagonist in rodents: evidence supporting the use of mGlu2/3 antagonists, for the treatment of depression. J Pharmacol Exp Ther 361:68–86
Witkin JM, Monn JA, Schoepp DD, Li X, Overshiner C, Mitchell SN, Carter G, Johnson B, Rasmussen K, Rorick-Kehn LM (2016) The rapidly acting antidepressant ketamine and the mGlu2/3 receptor antagonist LY341495 rapidly engage dopaminergic mood circuits. J Pharmacol Exp Ther 358:71–82
Witkin JM, Monn JA, Li J, Johnson B, McKinzie DL, Wang XS, Heinz BA, Li R, Ornstein PL, Smith SC, Mitch CH, Calligaro DO, Swanson S, Allen D, Phillips K, Gilmour G (2017b) Preclinical predictors that the orthosteric mGlu2/3 receptor antagonist LY3020371 will not engender ketamine-associated neurotoxic, motor, cognitive, subjective, or abuse-liability-related effects. Pharmacol Biochem Behav 155:43–55
Wright RA, Johnson BG, Zhang C, Salhoff C, Kingston AE, Calligaro DO, Monn JA, Schoepp DD, Marek GJ (2013) CNS distribution of metabotropic glutamate 2 and 3 receptors: transgenic mice and [^{3}H]LY459477 autoradiography. Neuropharmacology 66:89–98

Yang C, Shirayama Y, Zhang JC, Ren Q, Yao W, Ma M, Dong C, Hashimoto K (2015) R-ketamine: a rapid-onset and sustained antidepressant without psychotomimetic side effects. Transl Psychiatry 5:e632. https://doi.org/10.1038/tp.2015.136

Yoshimizu T, Shimazaki T, Ito A, Chaki S (2006) An mGluR2/3 antagonist, MGS0039, exerts antidepressant and anxiolytic effects in behavioral models in rats. Psychopharmacology 186:587–593

Zanos P, Highland JN, Stewart BW, Georgiou P, Jenne CE, Lovett J, Morris PJ, Thomas CJ, Moaddel R, Zarate CA Jr, Gould TD (2019) (2R,6R)-hydroxynorketamine exerts mGlu2 receptor-dependent antidepressant actions. Proc Natl Acad Sci U S A 116:6441–6450

Zanos P, Moaddel R, Morris PJ, Georgiou P, Fischell J, Elmer GI, Alkondon M, Yuan P, Pribut HJ, Singh NS, Dossou KS, Fang Y, Huang XP, Mayo CL, Wainer IW, Albuquerque EX, Thompson SM, Thomas CJ, Zarate CA Jr, Gould TD (2016) NMDAR inhibition-independent antidepressant actions of ketamine metabolites. Nature 533:481–486

Zhang K, Dong C, Fujita Y, Fujita A, Hashimoto K (2018) 5-Hydroxytryptamine-independent antidepressant actions of (R)-ketamine in a chronic social defeat stress model. Int J Neuropsychopharmacol 21:157–163

Zarate CA Jr, Singh JB, Carlson PJ, Brutsche NE, Ameli R, Luckenbaugh DA, Charney DS, Manji HK (2006) A randomized trial of an N-methyl-D-aspartate antagonist in treatment-resistant major depression. Arch Gen Psychiatry 63:856–864

Antidepressant Effects of the Muscarinic Receptor Antagonist Scopolamine: Clinical and Preclinical Review

Maura L. Furey, Wayne C. Drevets, and Anindya Bhattacharya

Abstract The nonselective muscarinic antagonist scopolamine is a rapid-acting antidepressant in limited clinical studies and shares common downstream signaling partners to ketamine, which is much more thoroughly studied both in the clinic and in preclinical models. In psychiatry, a clinically validated pathway is a rare phenomenon that holds particular importance for guiding neuroscientists toward neurobiological targets with the potential of affording more effective medicines for patients: either through reverse translation in search of targets that retain the clinical efficacy but increase the therapeutic window or through formulations that can be tailored to dosing schedules amenable to patients' ease of use. Scopolamine's rapid antidepressant profile provides an opportunity to deliver optimized medicines to patients, much like the recent clinical success of S-ketamine that ensued from robust follow-up investigation of ketamine's clinical profile. This chapter is an attempt to review the clinical and preclinical antidepressant data of scopolamine with a view on illuminating the critical muscarinic receptor subtype(s) that could prove fruitful as a next-generation antimuscarinic drug target for depression.

Keywords Scopolamine · Muscarinic receptor · Depression · Rapid acting

In the management of patients suffering from major depressive episodes, a great need persists for novel antidepressant therapies that are both rapidly acting and have a safety, tolerability, and ease of use profile to support widespread use. Over the past two decades, studies have demonstrated the ability of novel approaches – including the recent clinical success with racemic ketamine, S-ketamine (Berman et al. 2000; Zarate Jr. et al. 2006; Daly et al. 2018; Daly et al. 2019), and sleep deprivation

---

M. L. Furey
Neurocrine Bioscience, San Diego, CA, USA

W. C. Drevets · A. Bhattacharya (✉)
Janssen R&D, LLC. Neuroscience Therapeutic Area, San Diego, CA, USA
e-mail: abhatta2@its.jnj.com

K. Hashimoto, M. Manto (eds.), *New Rapid-acting Antidepressants*, Contemporary Clinical Neuroscience,
https://doi.org/10.1007/978-3-030-79790-4_8

therapy (Giedke and Schwarzler 2002; Shelton and Loosen 1993; Wu et al. 2009) – to provide symptom improvement within hours, in contrast to the 3–6 weeks required by conventional antidepressant pharmacotherapies. Here we provide an overview of the evidence showing that the nonselective muscarinic receptor antagonist, scopolamine, also produces antidepressant efficacy with relatively rapid onset of effect. The information reviewed herein begins with a series of randomized, controlled, clinical trials conducted in the National Institute of Mental Health (NIMH) Intramural Research Program and then summarizes work conducted by independent researchers. These clinical results are reviewed within the context of an earlier literature that hypothesized that the muscarinic cholinergic receptor system plays a major role in the pathophysiology of major depressive disorder (Dulawa and Janowsky 2019). These clinical findings encouraged efforts to elucidate the key muscarinic receptors in the brain that are critical to maintain the onset and duration of antidepressant efficacy provided by scopolamine while decreasing the burden on adverse events. An overview of these preclinical findings including a current status also is provided.

Muscarinic Cholinergic Antagonists in the Treatment of Mood Disorders

Interest in the muscarinic cholinergic system as a therapeutic target in mood disorders stemmed from evidence suggesting that hypersensitivity of the cholinergic system contributes to the underlying pathophysiology of mood disorders (Janowsky 1972). Increasing cholinergic activity through the inhibition of anticholinesterase by physostigmine provided a challenge uniquely capable both of rapidly exacerbating depressive symptoms in patients with major depressive disorder (MDD) and of switching the mood state in manic patients with bipolar disorder (BD) from euphoria to dysphoria (Janowsky 1972; Davis et al. 1978; Janowsky et al. 1974; Nurnberger Jr. et al. 1983; Risch et al. 1981). Cholinergic agonists including both arecoline (Sunderland et al. 1988) and oxotremorine (Davis et al. 1987) have been reported to worsen mood state, while critically, the depressogenic effects of physostigmine were blocked by the administration of scopolamine (Janowsky et al. 1986). Donepezil, which like physostigmine is an anticholinesterase, has been associated with elevated rates of relapse of major depressive disorder in elderly patients receiving this drug to manage symptoms of mild cognitive impairment (Reynolds 3rd et al. 2011). In addition, the cholinergically mediated neuroendocrine and pupillary responses to physostigmine (Dilsaver 1986; Janowsky et al. 1985; Rubin et al. 1999) were abnormally increased in individuals with MDD, consistent with a hypersensitive cholinergic response.

The muscarinic cholinergic receptor system was implicated more specifically by evidence that muscarinic receptor supersensitivity exists in depressed individuals (Dulawa and Janowsky 2019). For example, the onset of REM (rapid eye movement) as measured by polysomnography is advanced by a muscarinic cholinergic agonist (Riemann et al. 1994a), and this effect is exaggerated in patients with depression (Berger et al. 1989; Gillin et al. 1979). In addition, an earlier onset of REM in response to a muscarinic receptor agonist predicted subsequent onset of illness in individuals at risk for depression based on family history (Lauer et al. 2004). Some abnormalities in cholinergic receptor function in mood disorders showed sex effects. Specifically, sex differences were observed in the baseline and cholinergically stimulated plasma hormone measures that show differences between depressed and control groups, with results indicating that elevated cholinergic sensitivity occurs preferentially in premenopausal female MDD patients (Rubin et al. 1999; Rubin et al. 2003; Riemann et al. 1994b). Moreover, genetic variation in the CHRM2 gene (A/T 1890) reportedly was associated with MDD specifically in females (Comings et al. 2002). These observations are noteworthy in light of evidence reviewed below showing that women have larger antidepressant responses to scopolamine than men.

Collectively these data supported the hypothesis that antimuscarinic drugs may produce antidepressant effects and, based upon how rapidly physostigmine produced depressogenic outcomes, raised expectations that mood improvements may develop rapidly. Nevertheless, early studies reported modest and inconsistent antidepressant responses to antimuscarinic agents (Newhouse et al. 1988), although these were predominantly uncontrolled studies. In one open-label study, antidepressant effects were observed the day following an intramuscular administration of 0.4 mg of scopolamine (Gillin et al. 1991), but the magnitude of the effect was relatively small. The dose used in this study approximates half the bioequivalent dose later found to produce antidepressant effects, given the lower bioavailability associated with intramuscular administration relative to i.v. administration (Ebert et al. 2001) (notably the results of the dose-finding study presented below suggested that higher scopolamine doses were required to produce antidepressant responses of larger effect size). In another study, scopolamine was administered at doses up to 0.5 mg i.v. in patients with depression, and while no change in mood ratings (assessed using Profile of Mood States [POMS] depression factor scores) was observed by the study endpoint at 120 min post-infusion (Newhouse et al. 1988), follow-up assessments were not performed to evaluate the possibility that an emergent antidepressant effect may have developed over successive days. Subsequent studies using a similar i.v. dose (4.0 μg/kg) observed a similar clinical result acutely (i.e., no significant change in POMS depression factor ratings up to 150 min post-scopolamine administration) but also established the development of clinically meaningful antidepressant effects as early as 3 days following drug administration (Furey and Drevets 2006; Ketchum et al. 1973).

Rapid Antidepressant Response to Scopolamine

In a pilot study designed to assess the role of the muscarinic cholinergic system in the cognitive features in depression, Furey and Drevets (Furey and Drevets 2006) observed a rapid and robust antidepressant response to scopolamine in patients with MDD. Scopolamine was selected for its high affinity for all five muscarinic receptors, rapid entry into the brain, and slow dissociation rates from muscarinic receptors (Frey et al. 1992; Hyttel et al. 1985; Richelson 1983). In this pilot dose-ranging study, eight depressed individuals with MDD participated in four double-blind, randomly ordered testing sessions, in which participants received a 15-min i.v. infusion of saline placebo and each of three doses of scopolamine: 2.0, 3.0, and 4.0 μg/kg (Furey and Drevets 2006). The mean Montgomery-Asberg Depression Rating Scale (MADRS) scores obtained following 4.0 μg/kg of scopolamine were lower than both the baseline ($p = 0.0015$) and the pre-4.0 μg/kg measures ($p = 0.018$). Moreover, a larger reduction in MADRS scores pre- versus post-4.0 μg/kg of scopolamine than pre- versus post-placebo ($p = 0.01$) was observed, where post-assessments were obtained 3–5 days after the infusion. The mean change in MADRS score between the pretreatment baseline and the assessment following session 4 was -13.8 ± 7.7 ($p < 0.002$). Five of eight subjects met conventional criteria for response (>50% reduction in the MADRS score), and three of eight met criteria for remission (final MADRS score <10). The mean plasma concentrations for scopolamine increased with dose (mean ± SD = 82.8 ± 43.6, 122.1 ± 60.9, and 176.7 ± 71.3 ng/ml for the three scopolamine doses, respectively; $p < 0.001$). These promising observations were unexpected, however, and the pilot study was not designed to evaluate an antidepressant response. A second study thus was designed to test the antidepressant efficacy of scopolamine. Volunteers between 18 and 45 years of age were assessed for eligibility if they met DSM-IV-TR criteria for a current major depressive episode associated with either MDD or bipolar disorder (BD). Eighteen subjects (9 MDD, 9 BD) participated in a double-blind, placebo-controlled, crossover clinical trial consisting of multiple sessions in which they received i.v. infusions of placebo or scopolamine (4.0 μg/kg) administered over 15 minutes (Furey and Drevets 2006). All participants underwent a single-blind, placebo lead-in and then were randomized to either a P/S sequence or a S/P sequence, where P is a series of three placebo sessions and S is a series of three scopolamine sessions, each performed under double-blind conditions. The sessions were separated by 3–5 days. Clinical rating scales were acquired prior to each infusion. The primary outcome measure used to assess the antidepressant response was the change in MADRS scores. Secondary outcome measures included the Hamilton Anxiety Rating Scale (HAM-A) and POMS. The mean plasma concentrations of scopolamine did not differ significantly across the three 4.0 μg/kg scopolamine infusions (mean ± SD = 135.4 ± 74.3, 132.7 ± 77.5, and 106.8 ± 53.8, respectively).

Following completion of the baseline placebo infusion, the group randomized to receive scopolamine during study block one (S/P) showed a greater reduction in MADRS scores than the group randomized to receive placebo first (P/S); the placebo-adjusted (rate under scopolamine minus rate under placebo) reduction in MADRS scores under scopolamine was 52% ($p < 0.0001$; Cohen's $d = 2.7$). Similarly, in the P/S group, lower MADRS scores were observed in study block 2 (i.e., "S") as compared to both the baseline assessments ($p < 0.0001$; Cohen's $d = 3.2$) and study block 1 (i.e., "P"); here the placebo-adjusted reduction in MADRS scores under scopolamine was 66%; $p < 0.0001$, Cohen's $d = 3.4$. In both subgroups, clinical improvement was significant at the first assessment that followed scopolamine administration (i.e., 3–4 days after the initial administration), suggesting the antidepressant responses were relatively rapid. Crucially, most of the subjects who experienced a full clinical response to scopolamine (11/18 or 61%) also achieved remission (10/18) or 56%, regardless of whether they had a diagnosis of MDD or BD. The corresponding reductions on the HAM-A also were significant under scopolamine versus placebo ($p < 0.001$). Notably, the clinical improvement in depression and anxiety ratings experienced following scopolamine in the S/P group persisted through the subsequent placebo series, well beyond the expected duration of scopolamine's direct action at muscarinic receptors.

These clinical effects were replicated in an independent study that limited recruitment to depressed patients with recurrent MDD (Drevets and Furey 2009). Twenty-three depressed participants were randomized into the study, and 22 were included in the statistical analysis (one dropped out prior to receiving scopolamine). Using the double-blind, placebo-controlled, crossover design described above, participants were randomized into either the P/S ($n = 11$) or the S/P ($n = 11$) sequence. The group receiving scopolamine first (S/P) showed a 32% reduction in MADRS scores ($p < 0.001$) that exceeded the 6.5% decrease observed under placebo (P/S) ($p = 0.009$; Cohen's $d = 1.38$). Clinical improvement was significant at the first evaluation post-scopolamine administration ($p = 0.011$). The P/S group showed a 53% reduction in MADRS relative to baseline scores following the first scopolamine infusion ($p = 0.001$). As observed previously, the improvement seen during block 1 in the S/P group persisted through the placebo infusions in block 2 with no additional change in MADRS ($p = 0.53$). Sex differences in the antidepressant response magnitude following scopolamine administration were assessed in a separate study (Furey et al. 2010). A total of 52 (including 40 patients from the previously reported trials) male ($n = 21$) and female ($n = 31$)) outpatients meeting criteria for recurrent MDD or BD participated under the same experimental design described above. While the treatment effects as assessed by the MADRS were significant in males ($p = 0.043$) and females ($p < 0.001$) separately, an interaction with gender ($p = 0.009$) indicated a larger response magnitude in women. Similar effects were seen with the HAM-A, indicating that the anti-anxiety response to scopolamine also was greater in women ($p = 0.001$). Neither the AUC plasma concentration for scopolamine ($p > 0.20$) nor the drug-induced reductions in heart rate ($p > 0.20$) and blood pressure ($p > 0.20$) (pharmacodynamic effects that putatively are centrally mediated) differed between males and females (Furey et al. 2010). These data add to other literature showing sex effects with antidepressant treatments (Berlanga and Flores-Ramos 2006; Kornstein et al. 2000; Martenyi et al. 2001).

Adverse Effects

Importantly, the scopolamine i.v. dose range used for testing in mood disorders previously was shown to have effects on cognitive function but that did not have toxic effects such as delirium or psychosis (Newhouse et al. 1988). Compared to placebo, i.v. scopolamine administration (4.0 ug/kg) resulted in higher rates of drowsiness, blurred vision, dry mouth, and light-headedness. Because of the pulsed dosing regimen (i.e., three doses administered across a 2-week period), these adverse effects were sufficiently transient (resolving in 2–4 hours) and well-tolerated that no subject withdrew due to a side effect. No serious adverse event was encountered; in particular, no subject developed mania or hypomania, delirium, psychosis, overt confusion, clinically significant cardiovascular effects, or treatment-emergent suicidal ideation. This adverse event profile observed in the studies reported here is consistent with the dose selected for these studies (Mewaldt and Ghoneim 1979; Petersen 1977). As compared to placebo, the heart rate and the systolic and diastolic BP *decreased* following scopolamine infusion, consistent with known central effects on parasympathetic autonomic function. Importantly, no subject developed symptoms of hypotension or evidence of cardiovascular insufficiency. The generalizability of these reported side effects is limited, however, by the exclusion of participants who were elderly or who had clinically significant medical illnesses.

In considering scopolamine's effects on cognitive functioning, several studies have evaluated the acute effect of scopolamine on neuropsychological performance. Broad evidence exists that scopolamine impairs cognitive functions including attention, learning, and memory (Mewaldt and Ghoneim 1979; Petersen 1977; Hasselmo and Wyble 1997) and can produce delirium at higher doses (Ketchum et al. 1973). Importantly, symptoms of delirium were not observed at doses used to produce antidepressant effects (Newhouse et al. 1988; Mewaldt and Ghoneim 1979; Petersen 1977).

Onset and Duration of Scopolamine's Clinical Antidepressant Effects

The onset of scopolamine's antidepressant action was established by 3 days post-administration; however, no MADRS ratings were obtained between the initial infusion session and the next planned study infusion 3–5 days later. Nevertheless, patients reported anecdotally that the improvement experienced in depression severity generally was apparent by the first morning post-infusion (i.e., within 24 hours of drug exposure). In contrast, based upon the ratings obtained using the POMS (Drevets and Furey 2009), no improvement in mood was evident within 150 min of scopolamine infusion, and no further POMS ratings were obtained until the time of the subsequent infusion. Further studies thus are necessary to fully characterize the timing of the onset of scopolamine's antidepressant effect. Based upon the MADRS ratings obtained prior to subsequent infusion sessions, the observed antidepressant response persisted beyond the final scopolamine administration by at least 2 weeks. In participants who received scopolamine during block 1, the antidepressant effect persisted throughout the administration of placebo during block 2, and thus the effects lasted in the absence of scopolamine for at least 10 to 14 days (Furey and Drevets 2006; Drevets and Furey 2009). The persistence of the antidepressant effect was confirmed by demonstrating that depression ratings did not differ between the S/P and P/S groups in the final study block, when both groups had comparable levels of improvement relative to the pretreatment baseline. In both study groups, subjects showed improvement across the three scopolamine infusions, suggesting that repeated administrations provided additional benefit. Nevertheless, whether the persistence in the antidepressant response can be maintained over a longer time interval using repeated administrations of scopolamine (e.g., analogous to the treatment regimens developed to maintain the antidepressant effects of S-ketamine (Daly et al. 2018)) merits further testing.

Limitations of the Extant Data

The generalizability of these clinical findings obtained using IV scopolamine was limited by the relatively small sample sizes studied to date and the recruitment of study participants at a single site. In addition, the side effects following scopolamine administration may compromise the blind to drug versus placebo conditions. Nonetheless, the transient nature of the side effects may have partially preserved the double-blind given the primary outcome measure (MADRS score) was obtained 3–5 days following each infusion when subjects were free of side effects. Another design feature that mitigated the likelihood of functional un-blinding by side effects was that the placebo was also administered intravenously, potentially increasing the placebo effect, as evinced by subjective reports of sedative side effects in 59% of placebo-administered subjects.

Other Clinical Studies Using Repeated Scopolamine Administrations in Depression

Additional controlled clinical trials have been conducted to assess further the clinical efficacy of scopolamine, utilizing different routes of administration, study designs, or clinical populations. Oral scopolamine was used as an adjunct to citalopram in a double-blind, placebo-controlled outpatient trial. Forty patients with MDD and a Hamilton Depression score of $\geq$22 were randomly assigned to received placebo (n = 20) or oral scopolamine at 1 mg/day (n = 20) in combination with citalopram for a 6-week trial (Khajavi et al. 2012). The authors reported that compared to placebo, patients receiving scopolamine had a higher response rate at week 4 (65% vs 30%; $p = 0.027$) and higher remission rates at end of study (6 weeks) (65% vs 20%; $p = 0.004$), suggesting that scopolamine is an effective adjunct therapy when added to an SSRI to which MDD patients have shown an inadequate clinical response. In another small trial (Taub 2019), 14 patients with MDD and a MADRS score of ≥20 were randomized to receive either placebo ($n = 7$) or a combination of naltrexone and scopolamine ($n = 7$). Like scopolamine, naltrexone at this low dose has been reported to have antidepressant effects (Mischoulon et al. 2017), and this study was conducted to evaluate the efficacy and safety of this combination when administered orally as an adjunctive therapy to ongoing treatment in a 4-week trial. Patients were randomized to receive either placebo or active treatment which included a combination of scopolamine (sublingual administration 0.15 mg B.I.D. increased to a maximum of 0.5 mg/day) and naltrexone (1 mg B.I.D. increased to a maximum of 3 mg daily). The patients receiving active treatment showed a significantly larger reduction in MADRS at end of study compared to the placebo group ($p = 0.03$). Finally, one study conducted in patients with MDD with a more severe and treatment-refractory form of illness observed no significant antidepressant effect of scopolamine relative to placebo (Park et al. 2019). In this study, patients with MDD (mean MADRS = 33 ± 4) were randomized to receive an i.v. infusion of either placebo or scopolamine (0.04 mg/kg) using the procedures described above by Furey and Drevets. This study found no significant difference in change in MADRS between the placebo and treatment groups at end of study, suggesting that scopolamine may not be efficacious for a more treatment-resistant patient group (15 of 23 patients were treatment resistant) (Park et al. 2019). Importantly however, neither blood pressure nor heart rate changes were reported as side effects in this study, while these changes were prominent in the earlier infusion studies. Moreover, side effects generally were reported as minor in this study, while previous studies observed more significant effects on drowsiness and blurred vision under scopolamine (Furey and Drevets 2006). The absence of these side effects together with the lack of information on scopolamine plasma levels raises questions regarding whether differences existed in plasma exposure to scopolamine in this patient cohort relative to previous cohorts where there was greater evidence of more robust pharmacodynamic effects.

Advancing Scopolamine Through Reverse Translation

The clinical findings demonstrating robust antidepressant effects of the nonselective anticholinergic agent, scopolamine, have resulted in the reverse translation to preclinical work. This effort has focused on delineating the potential role of selective muscarinic subtypes in the antidepressant effects observed clinically. The elucidation of the key muscarinic receptors in the brain that are critical to antidepressant efficacy would allow for the development of more selective agents that also reduce the side effect burden. There are five muscarinic acetylcholine receptor (mAchR) subtypes, each encoded by five distinct genes (chrm1–5). The mAchRs are members of the G-protein coupled receptor (GPCR) superfamily A, and all five subtypes are expressed in the CNS to varying abundance (Kruse et al. 2014; Thiele 2013). While mAchR1 (M1), mAchR3 (M3), and mAchR5 (M5) couple primarily to $G\alpha_{q/11}$ leading to intracellular calcium increases following receptor activation, mAchR2 (M2) and mAchR4 (M4) couple to $G\alpha_{i/o}$, leading to inhibition of cAMP. Generally speaking, M1, M3, and M5 are postsynaptic GPCRs, while M2 and M4 are presynaptic autoreceptors modulating acetylcholine release; thus, while blocking M1, M3, and M5 will dampen muscarinic acetylcholine signaling, antagonism of M2 and M4 will increase the release of acetylcholine and thus activate the signal transduction cascade for both muscarinic and nicotinic acetylcholine receptors.

The mAchRs have critical physiological functions such as maintaining heart rate, smooth muscle contractility, glandular secretion, and CNS functions (including attention, learning and memory, sleep, arousal, addiction); these functions largely have been elucidated by studying knockout mice (Wess et al. 2007). Of particular interest in mood disorders is the key role of muscarinic cholinergic receptor signaling in the regulation of neuroplasticity changes, which partly involves the regulation of glutamate release in corticolimbic pathways (see below). In addition, the cholinergic projections originating in the basal forebrain play a major role in tuning the firing parameters of sensory cortical neurons so they become differentially sensitive to salient sensory signals (Holland 2007; Mesulam et al. 1983). The activity of these cholinergic projections is modulated by afferent input from the amygdala, hippocampus, and orbitofrontal cortex, which interact during associative learning (e.g., appetitive and aversive conditioning) to integrate limbic and sensory input during the assessment of stimulus valence and incentive value (Sarter et al. 2001). Altered activity of this network is thought to maintain the negative emotional processing biases in memory and attention that manifest in depressed patients (Victor et al. 2012). The key role of cholinergic transmission within the extended network may afford a circuitry in which muscarinic cholinergic receptor antagonism can modulate the emotional processing biases and reward learning deficits that underlie disturbances of emotional thought and behavior during depression. Consistent with this hypothesis, in patients with major depressive disorder, the antidepressant response to scopolamine was both predicted by an abnormal hemodynamic response of the amygdala to sad faces prior to treatment and correlated with a shift toward the normative pattern of response posttreatment (Szczepanik et al. 2016).

The role of specific mAchR subtypes in CNS physiology is not entirely clear, although there is a general appreciation of the role of cholinergic neuron clusters in CNS circuitry, and there is a consensus that muscarinic receptors in the CNS are potential drug targets in neurodegenerative and neuropsychiatric disorders. The challenge in the field of CNS muscarinic pharmacology stems from the lack of subselective in vivo tool compounds that can complement knockout mice phenotypes. Another complicating factor is that most neurons (cell bodies and terminals) express multiple mAChRs. Lack of subtype selective brain-penetrant muscarinic antagonists or allosteric modulators contributed to conflicting reports regarding the potential physiological and pathophysiological roles of specific mAChR subtypes in the CNS.

Knockout (KO) mice of chrm1–5 have contributed significantly to our basic understanding of the role and expression of the five mAchR subtypes in the brain (Thomsen et al. 2018). First, all five KO mice are viable and fertile with no obvious gross abnormalities either in behavior or neuromorphology. M1 was the first muscarinic receptor to be deleted. M1 is the most abundant muscarinic receptor in the brain with dense localization in the hippocampus, striatum, amygdala, and cortex. M1 also is expressed in abundance in peripheral tissues such as vascular smooth muscles, vas deferens, submaxillary gland, and sympathetic ganglia. Robust phenotypes were observed with M1 KO mice in pilocarpine-induced seizure, and M1 knockout mice exhibited increased locomotor activity. M1 KO mice did not display any significant deficits in sensory-motor gating, nociception, motor coordination, and anxiety-related behaviors. Several studies have provided insight on the role of M1 receptors in learning and memory. For example, M1 KO mice performed equally well as littermate controls in the Morris water maze task, while another study concluded that M1 receptors are not essential for memory formation in the hippocampus but are most likely involved in processes underlying memory consolidation. More importantly, scopolamine induced cognitive deficits in both wild-type and M1 KO mice, suggesting that M1 antagonism may not be a cognitive liability. Studies with the M2 KO mice revealed the importance of M2 on cardiac function. In the brain, M2 is expressed in the basal forebrain including the amygdala and cingulate cortices as well as the striatum but is also enriched in the cardiac tissue, which may be a reason for concern for the acute cardiovascular effects induced by scopolamine; it is fair to state that the long-term cardiac liabilities of scopolamine are unknown. Cardiac M2 is critical in driving the negative chronotropic effects following vagal stimulation or administration of muscarinic agonists. M2 and M4 KO mice were critical in our understanding of their role in presynaptic autoinhibitory mechanisms. M4 is postulated to be an important presynaptic autoreceptor in the CNS and is expressed on cholinergic interneurons and glutamatergic projection neurons in the stratum. M4 is also postulated to modulate GABA release in medium spiny neurons. M3 is enriched in peripheral gastrointestinal and bladder tissues and is generally expressed in low levels in the CNS (hippocampus, piriform cortex). M3 KO studies revealed an important role of these receptors in the hypothalamus in controlling appetite.

M5 is the least studied mAchR subtype, yet is expressed exclusively within the CNS, albeit at low levels and with regional selectivity. Selective expression and brain specificity of M5 make a case for understanding the role of this subtype in CNS pathophysiology. If brain-permeable M5 selective antagonists/modulators can be delivered with specificity, this may provide a CNS muscarinic drug target with fewer side effects than observed with the nonselective antimuscarinic scopolamine. M5 knockout mice revealed a critical role of these receptors in acetylcholine-induced vasorelaxation of small diameter cerebral blood vessels, due to the expression of M5 in endothelial cells of these vascular tissues. M5 expression is limited to ventral tegmental area (VTA) dopaminergic neurons which project to the nucleus accumbens, where M5 activation may modulate dopamine release. M5 is also expressed in the substantia nigra, hypothalamus, and CA2 region of the hippocampus. Using M5 KO mice, several recent studies point to the role of M5 receptors in modulating dopamine release from dopaminergic neurons of the midbrain. As such recent data has emerged with brain-penetrant M5 antagonists, complementing M5 KO studies and suggesting a possible role of M5 antagonism in reward processing and anhedonia relevant to mood disorders, addiction, and substance abuse disorders.

Scopolamine in Preclinical Models

Despite the limitations, despair behaviors in rodents have long been associated with depression-like symptoms in humans. Serotonin and norepinephrine transport inhibitors that show clinical efficacy reverse translate in animal models of forced swim test and tail suspension test, as assessed by a decrease in their immobility time. In the forced swim test model, scopolamine has been shown to decrease immobility in a dose-dependent manner (Martin et al. 2017; Witkin et al. 2014). The authors demonstrated scopolamine exposures in the brain (and plasma) and assumed that the efficacy observed was mediated by binding of scopolamine to mAchRs, although a direct target engagement readout from the same cohort of animals would have been informative. In the tail suspension test, both in acute setting and 24 h post-dose, a lower dose (25 μg/kg, i.p.) of scopolamine caused increases in biochemical endpoints (pAkt, pmTOR, pS6K) in rat frontal cortex (Voleti et al. 2013). Moreover, the authors demonstrated increased spine density and spine head diameter in distal segments of layer V neurons of prefrontal cortex, 24 h after scopolamine administration, suggesting synaptic rearrangement or synaptogenesis and possibly enhanced synaptic function. In line with this observation, scopolamine was efficacious in decreasing the immobility time (forced swim test) 24 h after a single injection, an effect that was mediated by enhanced glutamate signaling (Voleti et al. 2013), producing a set of findings that are highly similar to the reported effects of ketamine (Abdallah et al. 2015; Duman et al. 2016). Increased BDNF release also has been associated with scopolamine's mechanism of action (Ghosal et al. 2018). The long-lasting efficacy of scopolamine (and ketamine) is thought to be due to increases in synaptic plasticity, where changes at the level of synaptic proteins reorganize the synaptic connectivity and persist well beyond drug exposure (Hare et al. 2017). The role of mTOR and AMPA receptors in mediating the efficacy of scopolamine also was reported by an independent group (Dong et al. 2018). Since ketamine and scopolamine exert their cellular effects by different cell surface receptors converging on the AMPA activation pathway mediated by glutamate surge, that subthreshold doses of ketamine and scopolamine produced a synergy in terms of efficacy in the model of the forced swim test provided support for a common or overlapping mechanism of action (Monteggia and Kavalali 2013; Petryshen et al. 2016). Collectively these data unequivocally point to the robust efficacy of scopolamine in animal models of depression.

Identifying a More Selective Antimuscarinic Antidepressant

One of the important questions that remain is which receptor subtype (M1–M5) mediates the antidepressant action of scopolamine? More specifically, from a drug discovery point of view, what mAchR subtype in the brain maintains scopolamine's efficacy (and rapid onset) but spares the cardiac and GI side effects? Based on expression of mAchRs (see above), M2 and M3 are obvious receptor subtypes to avoid in a scopolamine-follow on antidepressant drug discovery program, leaving M1, M4, and M5 as potential targets. Given M4 is predominantly a presynaptic autoinhibitory receptor, M4 blockade would most likely induce a cholinergic drive and thus conceivably may precipitate mood disorder symptoms. In fact, M4 potentiation seems more in line with therapy for neuropsychiatric disorders. By elimination, M1 and M5 antagonism remain as the preferential targets for an antidepressant drug discovery program. In support of this position, scopolamine's efficacy was completely abolished in M1 KO mice and was partially absent in M5 knockout mice at a dose of 1 mg/kg (i.p.) (Witkin et al. 2014), although the authors did not focus on M5 in the publication. In addition, the expression pattern of M1 and M5 in the brain would support the M1 and/or M5 antagonism hypothesis, with M1 expression in the hippocampus/cortical layers and M5 expression supporting the mesolimbic dopamine reward circuit in depression. The elegant study by Wohleb et al. (Wohleb et al. 2016) demonstrated the role of M1 receptors in mediating the scopolamine effect on depression-like behaviors in preclinical models; they demonstrated M1 expression on cortical GABAergic interneurons (somatostatin positive). Together, these studies point to some role of M1 antagonism in mediating the clinical antidepressant efficacy of scopolamine. Whether M1 expression is limited to GABAergic interneurons only remains unclear, and if not, why selective knockdown of M1 in somatostatin GABAergic interneurons is enough to eradicate the efficacy of scopolamine will require further investigation. Nonetheless, M1 remains a feasible target, and M1 antagonism may be a mechanism worth pursuing in the clinic. The profile (how fast and how long) of M1 antagonism necessary to produce clinical efficacy will be somewhat critical, given the role of these receptors in memory consolidation.

The role of M5 in neuropsychiatry is more of a hypothesis currently, primarily driven by the unique expression pattern in the mesolimbic dopamine reward circuit, which has been implicated in the pathophysiology of mood disorders by multiple lines of evidence. Several studies have shown the role of stress in models of anhedonia where VTA dopaminergic neurons, via release of dopamine in the nucleus accumbens, contribute to the stress phenotype in animal models. As such it is plausible that a M5 selective antagonist may be efficacious in a stress model, as M5 is the sole mAchR in the VTA cell bodies. The specific cell type (dopaminergic, glutamatergic, GABAergic) within the VTA that expresses M5 receptors remains an active area of investigation. In summary, it is well accepted that scopolamine produces robust efficacy in animal models of depression, and the emerging hypothesis points to a M1 and/or M5 antagonism approach as a novel avenue to target depression, with a goal of maintaining the rapid onset and long-lasting duration of scopolamine observed in the clinic.

A selective antimuscarinic ligand is essential to drive the receptor subtype-mediated pharmacology, and drug discovery programs have not yet successfully brought negative modulators (NAM) of mAchRs to the clinic or to the market. With all five mAchRs now crystalized in an antagonist-bound state (latest being the M5 subtype), complemented with a plethora of mutagenesis studies of the five receptor subtypes, it is now clear that novel binding sites outside the conserved acetylcholine pocket must be exploited to drive selectivity. In fact, positive allosteric modulators of mAchR subtype have been described with acceptable pharmacology, with M1/4 PAMs demonstrating chemical tractability and promise as clinical candidates. Interestingly, NAMs for mAchRs have not been described in the literature, except in a few recent publications (Bender et al. 2019; Gentry et al. 2013) on M5 NAMs with less than ideal potency but suitable to test in rodent models of disease (addiction). Whether M1/M5 is the right target set for scopolamine's rapid antidepressant profile is yet to be fully explored; target validation of this hypothesis is ultimately dependent on finding subtype selective modulators/antagonists with desired in vivo properties through innovative medicinal chemistry approaches. In addition to attaining selectivity by targeting less conserved regions of the receptor (such an extracellular loops), an important aspect of allosterism is to maintain spatiotemporal signaling of the endogenous ligand (such as acetylcholine) in physiology and use the modulator to fine-tune the desired aspects of pathophysiology. This is extremely critical in acetylcholine-induced muscarinic receptor-mediated mood pathology hypothesis. Acetylcholine is an important neurotransmitter, and "fine tusning" of the affinity and/or efficacy of acetylcholine to desired mAchR subtype is the key to clinical success without adverse events.

Summary

The clinical studies reviewed here provide evidence that scopolamine exerts rapid and robust antidepressant effects. While scopolamine was relatively well-tolerated at the dose used, a more tolerable profile without sedation or changes in heart rate and blood pressure would be clinically beneficial for improving the benefit-to-risk ratio of antimuscarinic therapy for mood disorders. Thus, the identification of the muscarinic receptor subtype or subtypes that selectively mediate the clinical antidepressant response may create the opportunity for the development of an efficacious rapidly acting antidepressant with a highly tolerable side effect profile.

Financial Disclosures

Drs. Furey and Drevets are named as co-inventors on the use patent, "Scopolamine in the Treatment of Depression." Drs. Furey and Drevets have assigned their rights in the patent to the US government but will share a percentage of any royalties that may be received by the government. Drs. Furey, Drevets, and Bhattacharya are employees of Janssen Research & Development LLC, of Johnson & Johnson, and hold equity in Johnson & Johnson.

References

Berman RM, Cappiello A, Anand A, Oren DA, Heninger GR, Charney DS et al (2000) Antidepressant effects of ketamine in depressed patients. Biol Psychiatry 47(4):351–354

Zarate CA Jr, Singh JB, Carlson PJ, Brutsche NE, Ameli R, Luckenbaugh DA et al (2006) A randomized trial of an N-methyl-D-aspartate antagonist in treatment-resistant major depression. Arch Gen Psychiatry 63(8):856–864

Daly EJ, Singh JB, Fedgchin M, Cooper K, Lim P, Shelton RC et al (2018) Efficacy and safety of intranasal esketamine adjunctive to oral antidepressant therapy in treatment-resistant depression: a randomized clinical trial. JAMA Psychiat 75(2):139–148

Daly EJ, Trivedi MH, Janik A, Li H, Zhang Y, Li X et al (2019) Efficacy of esketamine nasal spray plus Oral antidepressant treatment for relapse prevention in patients with treatment-resistant depression: a randomized clinical trial. JAMA Psychiat

Giedke H, Schwarzler F (2002) Therapeutic use of sleep deprivation in depression. Sleep Med Rev 6(5):361–377

Shelton RC, Loosen PT (1993) Sleep deprivation accelerates the response to nortriptyline. Prog Neuro-Psychopharmacol Biol Psychiatry 17(1):113–123

Wu JC, Kelsoe JR, Schachat C, Bunney BG, DeModena A, Golshan S et al (2009) Rapid and sustained antidepressant response with sleep deprivation and chronotherapy in bipolar disorder. Biol Psychiatry 66(3):298–301

Dulawa SC, Janowsky DS (2019) Cholinergic regulation of mood: from basic and clinical studies to emerging therapeutics. Mol Psychiatry 24(5):694–709

Janowsky DS (1972) el-Yousef MK, Davis JM, Sekerke HJ. A cholinergic-adrenergic hypothesis of mania and depression. Lancet 2(7778):632–635

Davis JL, Gerbrandt LK, Cherkin A (1978) Retroactive amnesia induced in chicks by the proline analog L-baikiain, without EEG seizures or depression. Physiol Behav 21(4):653–658

Janowsky DS, el-Yousef MK, Davis JM (1974) Acetylcholine and depression. Psychosom Med 36(3):248–257

Nurnberger JI Jr, Jimerson DC, Simmons-Alling S, Tamminga C, Nadi NS, Lawrence D et al (1983) Behavioral, physiological, and neuroendocrine responses to arecoline in normal twins and "well state" bipolar patients. Psychiatry Res 9(3):191–200

Risch SC, Kalin NH, Janowsky DS (1981) Cholinergic challenges in affective illness: behavioral and neuroendocrine correlates. J Clin Psychopharmacol 1(4):186–192

Sunderland T, Tariot PN, Newhouse PA (1988) Differential responsivity of mood, behavior, and cognition to cholinergic agents in elderly neuropsychiatric populations. Brain Res 472(4):371–389

Davis KL, Hollander E, Davidson M, Davis BM, Mohs RC, Horvath TB (1987) Induction of depression with oxotremorine in patients with Alzheimer's disease. Am J Psychiatry 144(4):468–471

Janowsky DS, Risch SC, Kennedy B, Ziegler M, Huey L (1986) Central muscarinic effects of physostigmine on mood, cardiovascular function, pituitary and adrenal neuroendocrine release. Psychopharmacology 89(2):150–154

Reynolds CF 3rd, Butters MA, Lopez O, Pollock BG, Dew MA, Mulsant BH et al (2011) Maintenance treatment of depression in old age: a randomized, double-blind, placebo-controlled evaluation of the efficacy and safety of donepezil combined with antidepressant pharmacotherapy. Arch Gen Psychiatry 68(1):51–60

Dilsaver SC (1986) Pathophysiology of "cholinoceptor supersensitivity" in affective disorders. Biol Psychiatry 21(8–9):813–829

Janowsky DS, Risch SC, Huey LY, Kennedy B, Ziegler M (1985) Effects of physostigmine on pulse, blood pressure, and serum epinephrine levels. Am J Psychiatry 142(6):738–740

Rubin RT, O'Toole SM, Rhodes ME, Sekula LK, Czambel RK (1999) Hypothalamo-pituitary-adrenal cortical responses to low-dose physostigmine and arginine vasopressin administration: sex differences between major depressives and matched control subjects. Psychiatry Res 89(1):1–20

Riemann D, Hohagen F, Krieger S, Gann H, Muller WE, Olbrich R et al (1994a) Cholinergic REM induction test: muscarinic supersensitivity underlies polysomnographic findings in both depression and schizophrenia. J Psychiatr Res 28(3):195–210

Berger M, Riemann D, Hochli D, Spiegel R (1989) The cholinergic rapid eye movement sleep induction test with RS-86. State or trait marker of depression? Arch Gen Psychiatry 46(5):421–428

Gillin JC, Sitaram N, Duncan WC (1979) Muscarinic supersensitivity: a possible model for the sleep disturbance of primary depression? Psychiatry Res 1(1):17–22

Lauer CJ, Modell S, Schreiber W, Krieg JC, Holsboer F (2004) Prediction of the development of a first major depressive episode with a rapid eye movement sleep induction test using the cholinergic agonist RS 86. J Clin Psychopharmacol 24(3):356–357

Rubin RT, Abbasi SA, Rhodes ME, Czambel RK (2003) Growth hormone responses to low-dose physostigmine administration: functional sex differences (sexual diergism) between major depressives and matched controls. Psychol Med 33(4):655–665

Riemann D, Hohagen F, Bahro M, Berger M (1994b) Sleep in depression: the influence of age, gender and diagnostic subtype on baseline sleep and the cholinergic REM induction test with RS 86. Eur Arch Psychiatry Clin Neurosci 243(5):279–290

Comings DE, Wu S, Rostamkhani M, McGue M, Iacono WG, MacMurray JP (2002) Association of the muscarinic cholinergic 2 receptor (CHRM2) gene with major depression in women. Am J Med Genet 114(5):527–529

Newhouse PA, Sunderland T, Tariot PN, Weingartner H, Thompson K, Mellow AM et al (1988) The effects of acute scopolamine in geriatric depression. Arch Gen Psychiatry 45(10):906–912

Gillin JC, Sutton L, Ruiz C, Darko D, Golshan S, Risch SC et al (1991) The effects of scopolamine on sleep and mood in depressed patients with a history of alcoholism and a normal comparison group. Biol Psychiatry 30(2):157–169

Ebert U, Grossmann M, Oertel R, Gramatte T, Kirch W (2001) Pharmacokinetic-pharmacodynamic modeling of the electroencephalogram effects of scopolamine in healthy volunteers. J Clin Pharmacol 41(1):51–60

Furey ML, Drevets WC (2006) Antidepressant efficacy of the antimuscarinic drug scopolamine: a randomized, placebo-controlled clinical trial. Arch Gen Psychiatry 63(10):1121–1129

Frey KA, Koeppe RA, Mulholland GK, Jewett D, Hichwa R, Ehrenkaufer RL et al (1992) In vivo muscarinic cholinergic receptor imaging in human brain with [11C]scopolamine and positron emission tomography. J Cereb Blood Flow Metab 12(1):147–154

Hyttel J, Larsen JJ, Christensen AV, Arnt J (1985) Receptor-binding profiles of neuroleptics. Psychopharmacology Suppl 2:9–18

Richelson E (1983) Antimuscarinic and other receptor-blocking properties of antidepressants. Mayo Clin Proc 58(1):40–46

Drevets WC, Furey ML (2009) Replication of scopolamine's antidepressant efficacy in major depressive disorder: a randomized, placebo-controlled clinical trial. Biol Psychiatry 67(5):432–438

Furey ML, Khanna A, Hoffman EM, Drevets WC (2010) Scopolamine produces larger antidepressant and antianxiety effects in women than in men. Neuropsychopharmacology 35(12):2479–2488

Berlanga C, Flores-Ramos M (2006) Different gender response to serotonergic and noradrenergic antidepressants. A comparative study of the efficacy of citalopram and reboxetine. J Affect Disord 95(1–3):119–123

Kornstein SG, Schatzberg AF, Thase ME, Yonkers KA, McCullough JP, Keitner GI et al (2000) Gender differences in treatment response to sertraline versus imipramine in chronic depression. Am J Psychiatry 157(9):1445–1452

Martenyi F, Dossenbach M, Mraz K, Metcalfe S (2001) Gender differences in the efficacy of fluoxetine and maprotiline in depressed patients: a double-blind trial of antidepressants with serotonergic or norepinephrinergic reuptake inhibition profile. Eur Neuropsychopharmacol 11(3):227–232

Mewaldt SP, Ghoneim MM (1979) The effects and interactions of scopolamine, physostigmine and methamphetamine on human memory. Pharmacol Biochem Behav 10(2):205–210

Petersen RC (1977) Scopolamine induced learning failures in man. Psychopharmacology 52(3):283–289

Hasselmo ME, Wyble BP (1997) Free recall and recognition in a network model of the hippocampus: simulating effects of scopolamine on human memory function. Behav Brain Res 89(1–2):1–34

Ketchum JS, Sidell FR, Crowell EB Jr, Aghajanian GK, Hayes AH Jr (1973) Atropine, scopolamine, and ditran: comparative pharmacology and antagonists in man. Psychopharmacologia 28(2):121–145

Khajavi D, Farokhnia M, Modabbernia A, Ashrafi M, Abbasi SH, Tabrizi M et al (2012) Oral scopolamine augmentation in moderate to severe major depressive disorder: a randomized, double-blind, placebo-controlled study. J Clin Psychiatry 73(11):1428–1433

Taub N (2019) Naltrexone and scopolamine rapidly reduce symptoms of major depressive disorder (MDD): a double blinded randomized controlled pilot study. Open J Depression 08:1–4

Mischoulon D, Hylek L, Yeung AS, Clain AJ, Baer L, Cusin C et al (2017) Randomized, proof-of-concept trial of low dose naltrexone for patients with breakthrough symptoms of major depressive disorder on antidepressants. J Affect Disord 208:6–14

Park L, Furey M, Nugent AC, Farmer C, Ellis J, Szczepanik J et al (2019) Neurophysiological changes associated with antidepressant response to ketamine not observed in a negative trial of scopolamine in major depressive disorder. Int J Neuropsychopharmacol 22(1):10–18

Kruse AC, Kobilka BK, Gautam D, Sexton PM, Christopoulos A, Wess J (2014) Muscarinic acetylcholine receptors: novel opportunities for drug development. Nat Rev Drug Discov 13(7):549–560

Thiele A (2013) Muscarinic signaling in the brain. Annu Rev Neurosci 36:271–294

Wess J, Eglen RM, Gautam D (2007) Muscarinic acetylcholine receptors: mutant mice provide new insights for drug development. Nat Rev Drug Discov 6(9):721–733

Holland PC (2007) Disconnection of the amygdala central nucleus and the substantia innominata/nucleus basalis magnocellularis disrupts performance in a sustained attention task. Behav Neurosci 121(1):80–89

Mesulam MM, Mufson EJ, Levey AI, Wainer BH (1983) Cholinergic innervation of cortex by the basal forebrain: cytochemistry and cortical connections of the septal area, diagonal band nuclei, nucleus basalis (substantia innominata), and hypothalamus in the rhesus monkey. J Comp Neurol 214(2):170–197

Sarter M, Givens B, Bruno JP (2001) The cognitive neuroscience of sustained attention: where top-down meets bottom-up. Brain Res Brain Res Rev 35(2):146–160

Victor TA, Furey ML, Fromm SJ, Bellgowan PS, Ohman A, Drevets WC (2012) The extended functional neuroanatomy of emotional processing biases for masked faces in major depressive disorder. PLoS One 7(10):e46439

Szczepanik J, Nugent AC, Drevets WC, Khanna A, Zarate CA Jr, Furey ML (2016) Amygdala response to explicit sad face stimuli at baseline predicts antidepressant treatment response to scopolamine in major depressive disorder. Psychiatry Res Neuroimaging 254:67–73

Thomsen M, Sorensen G, Dencker D (2018) Physiological roles of CNS muscarinic receptors gained from knockout mice. Neuropharmacology 136(Pt C):411–420

Martin AE, Schober DA, Nikolayev A, Tolstikov VV, Anderson WH, Higgs RE et al (2017) Further evaluation of mechanisms associated with the antidepressantlike signature of scopolamine in mice. CNS Neurol Disord Drug Targets 16(4):492–500

Witkin JM, Overshiner C, Li X, Catlow JT, Wishart GN, Schober DA et al (2014) M1 and m2 muscarinic receptor subtypes regulate antidepressant-like effects of the rapidly acting antidepressant scopolamine. J Pharmacol Exp Ther 351(2):448–456

Voleti B, Navarria A, Liu RJ, Banasr M, Li N, Terwilliger R et al (2013) Scopolamine rapidly increases mammalian target of rapamycin complex 1 signaling, synaptogenesis, and antidepressant behavioral responses. Biol Psychiatry 74(10):742–749

Abdallah CG, Sanacora G, Duman RS, Krystal JH (2015) Ketamine and rapid-acting antidepressants: a window into a new neurobiology for mood disorder therapeutics. Annu Rev Med 66:509–523

Duman RS, Aghajanian GK, Sanacora G, Krystal JH (2016) Synaptic plasticity and depression: new insights from stress and rapid-acting antidepressants. Nat Med 22(3):238–249

Ghosal S, Bang E, Yue W, Hare BD, Lepack AE, Girgenti MJ et al (2018) Activity-dependent brain-derived neurotrophic factor release is required for the rapid antidepressant actions of scopolamine. Biol Psychiatry 83(1):29–37

Hare BD, Ghosal S, Duman RS (2017) Rapid acting antidepressants in chronic stress models: molecular and cellular mechanisms. Chronic Stress (Thousand Oaks) 1

Dong J, Zhou Q, Wei Z, Yan S, Sun F, Cai X (2018) Protein kinase A mediates scopolamine-induced mTOR activation and an antidepressant response. J Affect Disord 227:633–642

Monteggia LM, Kavalali ET (2013) Scopolamine and ketamine: evidence of convergence? Biol Psychiatry 74(10):712–713

Petryshen TL, Lewis MC, Dennehy KA, Garza JC, Fava M (2016) Antidepressant-like effect of low dose ketamine and scopolamine co-treatment in mice. Neurosci Lett 620:70–73

Wohleb ES, Wu M, Gerhard DM, Taylor SR, Picciotto MR, Alreja M et al (2016) GABA interneurons mediate the rapid antidepressant-like effects of scopolamine. J Clin Invest 126(7):2482–2494

Bender AM, Garrison AT, Lindsley CW (2019) The muscarinic acetylcholine receptor M5: therapeutic implications and allosteric modulation. ACS Chem Neurosci 10(3):1025–1034

Gentry PR, Kokubo M, Bridges TM, Kett NR, Harp JM, Cho HP et al (2013) Discovery of the first M5-selective and CNS penetrant negative allosteric modulator (NAM) of a muscarinic acetylcholine receptor: (S)-9b-(4-chlorophenyl)-1-(3,4-difluorobenzoyl)-2,3-dihydro-1H-imidazo[2,1-a]isoi ndol-5(9bH)-one (ML375). J Med Chem 56(22):9351–9355

Index

K. Hashimoto, M. Manto (eds.), *New Rapid-acting Antidepressants*, Contemporary Clinical Neuroscience, https://doi.org/10.1007/978-3-030-79790-4

The manufacturer's authorised representative in the EU is Springer Nature Customer Service Centre GmbH, Europaplatz 3, 69115 Heidelberg, Germany. If you have any concerns regarding our products, please contact ProductSafety@springernature.com

Printed and bound by CPI Group (UK) Ltd, Croydon, CR0 4YY

30/09/2026

02225515-0006